T0328942

Non-governmental organizations and the sustainability of small and medium-sized enterprises in Peru

Non-governmental organizations and the sustainability of small and medium-sized enterprises in Peru

An analysis of networks and discourses

Walter V. Castro Aponte

Environmental Policy Series – Volume 9

Wageningen Academic
Publishers

ISBN: 978-90-8686-229-0
e-ISBN: 978-90-8686-783-7
DOI: 10.3920/978-90-8686-783-7

First published, 2013

© Wageningen Academic Publishers
The Netherlands, 2013

Preface

The day that I went to the Netherlands to study my MSc, I promised myself to return to my country with the highest academic degree: a PhD. So, after obtaining my MSc degree, I started to develop my PhD research proposal. My main motivation to pursue a PhD study on environmental policy has been the need to deepen my understanding of the patterns and perspectives that enable and constrain governing the environment, particularly regarding the role of non-state actors. The complex Peruvian reality and the failing of the Peruvian state in leading environmental improvement provided a rich context to study non-state actors and their roles towards sustainability.

This thesis deals with two groups of actors that are less investigated regarding their current and potential roles in making production and the market sustainable: NGOs, and the most rapidly growing and environmentally sensitive segment of Peru's national economy: SMEs. The changing roles of NGOs in promoting sustainability of SMEs in Peru are investigated through the perspectives of networks and discourses. The thesis investigates the actual and potential changes in networks and discourses that NGOs supporting SMEs are undergoing. The investigation focuses on the domains of organic production, business social responsibility and sustainable production using the theory of network society and the theory of ecological modernization as frameworks.

This PhD research project came out based on my personal experience, concern and commitment to the Peruvian people. The collaboration of Wageningen University in The Netherlands and the Pacifico University in Peru, under the financial sponsorship of the Wageningen University Sandwich PhD Fellowship, has made this achievement feasible. I am very grateful to them for believing in the value of my PhD research project. During the past years, many people have contributed in different ways to my research. I owe very special thanks to my major professor and promoter, Prof. dr. ir. Arthur P.J. Mol, Chair of the Environmental Policy (ENP) group at Wageningen University, The Netherlands, and to my co-promoter, dr. ir. Kris van Koppen, associate professor of the ENP group at Wageningen University, the Netherlands. Both professors guided me throughout the entire research, from the start by commenting my early writing drafts until the end by editing the dissertation and the propositions. Dr. ir. Kris van Koppen guided me in the details of preparing the research proposal, helped me to design and plan the research, and assisted me to solving most of the difficulties faced while the research was being conducted. His extensive knowledge of environmental issues, policy and management was reflected by his critical comments to the dissertation drafts. Words are not sufficient to express my gratitude to him.

I express my deep appreciation to Dr. Winfried Timmers, Director of Tactum company. He was the one who encouraged me in elaborating my own PhD research proposal. I could develop the research proposal, implement the research and finalize successfully the PhD thesis primarily thanks to his trust. He always cared about, and encouraged me during, the critical moments of my research.

Thanks are due to Dr. Eduardo Moron, Director of the Centro de Investigación de la Universidad del Pacifico (CIUP) in Peru. Thanks are also due to Dr. Carlos Loret de Mola de Lavalle and Mariano Castro Sánchez Moreno, president and executive secretariat of the National Environmental Council (CONAM in Spanish acronym) respectively. Their support was key to starting the PhD research. Special thank to Elsa Galarza, Rosario Gomez, Felipe Portocarrero and Cynthia Sanborn, researchers at CIUP, for their support and guidance during the fieldwork period of the research.

I would like to express my thanks to the local and international NGOs, local SMEs and other organizations/institutions mentioned in the case studies for their kind collaboration and contributions to my research data.

While studying and doing my research at the Wageningen University's ENP group, I enjoyed very much, thanks to the kind assistance provided by all the ENP staff to whom I am grateful. Many thanks go to Dr. ir. Peter Oosterveer, Prof. dr. ir. Gert Spaargaren and Prof. dr. ir. Jan van Tatenhove for providing me additional literature and encouragement.

My life and study in Wageningen was so much interesting thanks to the great friendship provided by the PhD students at the ENP group. I would like to thank Judith van Leeuwen, Jorrit Nijhuis, Michiel de Krom, Dries Hegger, Elizabeth Sargant and Maria Tysiachniouk. I would also like to extend my thanks to others, whose names are not listed here but who have also, directly or indirectly, helped me in various ways by contributing to this dissertation.

My life in Wageningen has been unforgettable. This is largely due to the kind friendships provided by a number of people. My gratitude to my unforgettable friends Sander ten Meulen, Jordi ter Meulen, Raul van Veek, Peter van der Werf, Jonah van Beijnen, Maarten van Soest, Hugo Calvillo, Nico Mentik, Enrique Diaz and Claudia Pavon. Their companionship taught me the value of trust and intercultural friendship. My deepest gratitude also to Hein Schafrat, Marta Evelina Collahuacho, Doris Collahuacho, Alma Inkari, Juana Vera and Zamir. They have encouraged me to conduct and finalize this research and their thoughts always were a source of fresh ideas that enriched my understanding of Peru and Europe.

Finally, I dedicate this work to my mother and my father who, during their entire lives, encourage their children to become well educated. I thank my sisters and brother for strongly supporting and continuously encouraging me during my higher education. Last but not least, I thank Edith (Kusiquyllur) Casafranca, my *yanallay* for her love, trust and patience, especially during the writing stage of the thesis. I devote this work to Qarwapuma Rikra, my one year old son. I owe them the rest of my life. In memory of my grandparents, examples of living in harmony with nature.

Wageningen May, 2013

Table of contents

Abbreviations

ADEPIA	Asociación de Empresas del Parque Industrial de Arequipa (Industrial Park of Arequipa Associations)
ADEX	Asociación de Exportadores del Perú (Peruvian Exporters Association)
ADG	Aid for Development Gembloux
AECI	Agencia Española de Cooperación Internacional (Spanish Agency for International Cooperation)
AEDES	Asociación Especializada para el Desarrollo Sostenible (Specialized Association for Sustainable Development)
AIAB	Italian Association of Biological Agriculture
ANPE	Asociación Nacional de Productores Ecológicos (National Ecological Producers Association of Peru)
AOPEB	Asociación de Organizaciones de Productores Ecológicos de Bolivia (Association of ecological producer organizations of Bolivia)
APEC	Asia-Pacific Economic Cooperation
APEGA	Sociedad Peruana de Gastronomía (Peruvian Society of Gastronomy)
APEPA	Asociación de Productores Ecológicos de Plantas Aromáticas de Sihuas-Ancash (Ecological Producer Association of Aromatic Plants of Sihuas-Ancash)
APETRICES	Asociación Peruana de Productores de Trigo, Cebada y Sucedáneos (Peruvian Association of Wheat and Barley)
APOMIPE	Programa de Apoyo a la Micro y Pequeña Empresa en el Perú (Program of Support to Micro and Small Enterprises in Peru)
APPEAP	Asociación de Productores Ecológicos del Alto Piura (Association of Ecological Producers of Alto Piura)
APROMALPI	Asociación de Productores de Mango del Alto Piura (Association of Producers of Mango of Alto Piura)
ARPEP	Asociación Regional de Productores Orgánicos de Piura (Regional Association of Organic Producers of Piura)
ARPOA	Asociación Regional de Productores Orgánicos de Ayacucho (Regional Organic Producer Association of Ayacucho)
ASPEC	Asociación Peruana de Consumidores y Usuarios (Peruvian Association of Consumers and Users)
CAB	Convenio Andres Bello (Andres Bello Intergovernmental Treaty)
CACYT	Consejo Andino de Ciencia y Tecnología (Andean Council of Science and Technology)
CAN	Comunidad Andina de Naciones (Andean Community of Nations)
CBOs	Organizaciones de base (Community-Based Organizations)
CCE	Comité de Consumidores Ecológicos (Committee of Ecological Consumers)
CCP	Confederación Campesina del Perú (Peasants Confederation of Peru)

CDM	Clean Development Mechanism
CEAR	Centro de Apoyo Rural (Rural Support Center)
CEDAL	Centro de Asesoría Laboral del Perú (Labor Advisory Council of Peru)
CEDECAP	Centro de Demostración y Capacitación en Tecnologías Apropiadas (Centre of Demonstration and Qualification in Appropriate Technologies)
CEDEP	Centro de Estudios para el Desarrollo y la Participación (Center of Studies for Development and Participation)
Centro IDEAS	Centro de Investigación, Documentacion, Educación. Asesoría y Servicios (Center for Research, Documentation, Education, Advice and Services)
CEPAL	Economic Commission for Latin America and the Caribbean (Comisión Económica para América Latina y el Caribe)
CEPIBO	Central Piurana de Asociaciones de Pequeños Productores de Banano Orgánico (Association of Small-Scale Producers of Organic Bananas from Piura)
CEPICAFE	Central Piurana de Cafetaleros (Piura Coffee Growers Association)
CER	Centro de Ecoeficiencia y Responsabilidad Social (Eco-efficiency and Social Responsibility Center)
CESEM	Centro de Servicios Empresariales (Service Center for Businesses of Arequipa)
CFAs	Dutch Co-Financing Agencies
CI	Consumers International
CID	Programa de Competitividad, Innovación y Desarrollo de la Región Arequipa (Program of Competitiveness, Innovation and Development of the Arequipa Region)
CIP	Centro Internacional de la Papa (International Potato Center)
CITEs	Centros de Innovación tecnológica (Centers of Technological Innovation)
CNA	Confederación Nacional Agraria (National Agrarian Confederation)
COCLA	Central de Cooperativas Agrarias Cafetaleras (Center of Coffee Grower Cooperatives)
CONACS	Consejo Nacional de Camélidos Sudamericanos (National Council for South-American Camelids)
CONAM	Consejo Nacional del Ambiente (National Council of Environment)
CONAPO	Comisión Nacional de Productos Orgánicos (National Commission of Organic Products)
New CONAPO	Consejo Nacional de Productos Orgánicos (National Council of Organic Products)
CONCYTEC	Consejo Nacional de Ciencia y Tecnología del Perú (National Council of Science and Technology of Peru)
CONDESAN	Consorcio para el Desarrollo Sostenible de la Ecoregión Andina (Consortium for the Sustainable Development of Andean Ecorregion)
CONVEAGRO	Convención Nacional del Agro Peruano (National Convention for Peruvian Agriculture)

COPEME	El Consorcio de Organizaciones Privadas de Promoción al Desarrollo de la Pequeña y Microempresa (Peruvian Consortium of Private Organizations for the Promotion of Small and Medium-sized Business Development)
CORDAID	Catholic Organization for Relief and Development AID
CPLatinNet	Red Latinoamericana de Producción Más Limpia (Latin American Cleaner Production Platform)
CRTT	Center for Technological Transferring Resources
CYTED	Programa Iberoamericano de Ciencia y Tecnología para el Desarrollo (Iberoamerican Program of Science and Technology for Development)
DANIDA	Danish International Development Agency
DARCOF	Danish Research Center for Organic Farming
DED	German Development Service
DGCA	Dirección General de Competitividad Agraria del Ministerio de Agricultura (Ministry of Agriculture's General Direction of Agrarian Competitiveness)
DIGESA	Dirección Nacional de Salud Ambiental (National Direction of Environmental Health)
EAP	Economically Active Population
ECODES	Ecología y Desarrollo (Ecology and Development)
EMPA	Swiss Federal Laboratories for Materials Testing and Research
EMSs	Environmental Management Systems
ENPEs	Encuentro Nacional de Productores Ecológicos del Perú (National Meeting of Ecological Producers)
ETI	Ethical Trading Initiative
EU	European Union
FAO	Organization of the United Nations
FiBL	Research Institute of Organic Agriculture
FLA	Fair Labor Association
FONCODES	Fondo de Cooperación para el Desarrollo Social (Cooperation Fund for Social Development)
FOVIDA	Fomento para la Vida (Support for Life)
GALCI	Grupo de América Latina y el Caribe del IFOAM (IFOAM Group for Latin America and the Caribbean)
GAP	Good Agricultural Practices
GDP	Gross Domestic Product
GEF	Global Environmental Facility
GHG	Greenhouse gas
GM	Genetic Modified
GNI	Gross National Income
GRESP	Red Peruana para la Economía Solidaria (Peruvian Network of Solidarity Economy)
Grupo GEA	Grupo de Emprendimientos Ambientales (Environmental Entrepreneurship Group)

GTZ	German Technical Co-operation
HIVOS	Humanist Institute for Cooperation with Developing Countries
ICCO	Interchurch Organization for Development Cooperation
ICS	Internal Control System
IDB	Inter-American Development Bank
IDMA	Instituto de Desarrollo y Medio Ambiente (Institute of Development and Environment)
IDRC	International Development Research Center
IFAD	International Fund for Agriculture
IFOAM	International Federation of Organic Agriculture Movements
ILO	International Labour Organization
ICT	Information and Communication Technologies
INIA	Instituto Nacional de Investigación Agraria (National Institute for Agrarian Innovation)
INOFO	Intercontinental Network of Organic Farmers Organizations
IPES	Instituto Peruano de Economía Social (Peruvian Institute of Social Economy)
ITACAB	Instituto de Transferencia de Tecnologías Apropiadas para Sectores Marginales (Institute for the Transfer of Technology for Marginal Sectors)
ITDG	Intermediate Technology Development Group
JNC	Junta Nacional del Café (National Union for Coffee Producers)
MAELA	Movimiento Agroecológico de América Latina (Latin American Agroecological Movement)
MAOCO	Movimiento de Agricultura Orgánica Costarricense (Costa Rican Organic Agriculture Movement)
MDGs	Millennium Development Goals
MEF	Ministerio de Economía y Finanza del Perú (Ministry of Economy and Finance of Peru)
MIF	Multilateral Investment Fund (Fondo Multilateral de Inversiones: FOMIN)
MINAG	Ministerio de Agricultura del Perú (Ministry of Agriculture of Peru)
MINAM	Ministerio de Ambiente del Perú (Ministry of Environment of Peru)
MINCETUR	Ministerio de Comercio Exterior y Turismo del Perú (Ministry of Foreign Commerce and Tourism of Peru)
MISEREOR	German catholic bishops' organization for development cooperation
MYPE	Micro y Pequeñas Empresas (Small and Micro-sized Enterprises)
NCPCs	Centros Nacionales de Producción Más Limpia (National Cleaner Production Centers)
NGOs	Non-Governmental Organizations
OAS	Organization of American States
OECD	Organisation for Economic Co-operation and Development
ONUDI	United Nations Industrial Development Organization
OTCIT	Oficina Central de las CITEs (Central Office of CITEs)

PAM	Movimiento Agroecológico Peruano (Peruvian Agroecological Movement)
PCM	Consejo de Ministros del Perú (Ministries' Council of Peru)
PGS	Participatory Guaranty Systems
PIDAASSA PERU	Programa de Intercambio, Diálogo y Asesoría en Agricultura Sostenible y Seguridad Alimentaria (Program for Exchange, Dialogue and Consultation on Sustainable Agriculture and Food Sovereignty)
PIDECAFE	Programa Integral para el Desarrollo del Café (Integrated Programme for Coffee Development)
POPs	Persistent Organic Pollutants
PRODUCE	Ministerio de Producción del Perú (Ministry of Production of Peru)
PROMPERU	Comisión de Promoción del Perú para la Exportación y el Turismo (Commission on the Promotion of Peru for Export and Tourism)
PROMPYME	Comisión para la Promoción de la Pequeña y Micro-empresa (Center for the Promotion of Small and Micro Enterprises)
R&D	Research and Development
RAAA	Red Peruana para la Acción en Agricultura Alternativa (Peruvian Network for Action in Alternative Agriculture)
RAE Perú	Red de Agricultura Ecológica del Perú (Ecological Agriculture Network of Peru)
RECP	Joint UNIDO-UNEP Programme on Resource Efficiency and Cleaner Production
REDAR	Red de Agroindustria Rural del Perú (Network of Rural Agri-industries of Peru)
REMURPE	Red de Municipalidades Rurales del Perú (Network of Rural Municipalities of Peru)
REPEBAN	Organización de Productores de Banano Orgánico (Network of Organic Banana Producers)
RESIRDES	Gestión de la evaluación, seguimiento e introducción de resultados de la ciencia, la tecnología y de la innovación tecnológica para incrementar su impacto en el desarrollo económico y la competitividad (Evaluation, monitoring of science, technology and innovation for improving economic development and competitiveness)
SAI	Social Accountability International
SANET	Sustainable Alternatives Network
SDS	Swiss international Cooperation Agency
SECO	State Secretariat for Economic Affairs of the Swiss Federation
SENASA	Servicio Nacional de Sanidad Agraria (National Service of Agrarian Health)
SENATI	Servicio Nacional de Adiestramiento en Trabajo Industrial (National School for Industrial Training)
SHP	Sociedad de Hoteles del Perú (Hotels Society of Peru)
SMEs	Small and Medium-sized Enterprises

SNA	Sociedad Nacional del Ambiente (National Environmental Society)
SNI	Sociedad Nacional de Industria (National Society of Industry)
TLC	Tratado de Libre Comercio (Free Trade Agreement)
TTN	Technological Transfer Network
UNALM	Universidad Nacional Agraria La Molina (National Agrarian University)
UNCED	United Nations Conference on Environment and Development
UNDP	United Nations Development Programme
UNEP	United Nations Environmental Programme (Programa de las Naciones para el Medio Ambiente: PNUMA)
UNIDO	United Nations Industrial Development Organization
UPAS	Unidades Productivas Agropecuarias
USAID	United States Agency for International Development
WASTE	Dutch Consultancy
WBCSD	World Business Council for Sustainable Development
WRI	World Resource Institute
WSSD	World Summit for Sustainable Development
WTO	World Trade Organization

Chapter 1.
Introduction

1.1 Background and problem description

Small and medium-sized enterprises (SMEs) have been recognized as a major source of employment and income in developed and developing countries (Bridge, O'Neill & Martin, 2009; Mead & Liedholm, 1998; Pradhan, 1989). Roughly 70% of global economic activities are generated by SMEs (Cerin, 2004). In Latin American countries SMEs play a substantial role in the development process (Peres & Stumpo, 2000). In the region, SMEs are important in economic and social development due to their contribution to the strategies of poverty alleviation and diminishing of unemployment rates (Corral, Isusi, Peinado-Vara & Pérez, 2005). The number of micro scale enterprises in Latin America, representing the majority of SMEs, is about 98% of the total number of enterprises (Ferraro & Stumpo, 2010). In the Andean countries SMEs represent more than 75% of the total number of companies and generate most of the employment (Flores, 2004; TTN-Red Andina, 2005).

In Peru the importance of SMEs for local and global economic supply chains is out of the question. According to Su (2004), Peruvian SMEs represent 98% of the total number of companies registered in the country. Small and micro scale enterprises provide employment to 53% of the economically active population (EAP) at the national level and 76% in Lima, the capital of the country (Corral *et al.*, 2005; Minaya, 2007). According to Arbulú (2007) small and micro-enterprises, the largest portion of SMEs in Peru, are responsible of 60% of the national employment. The number of Peruvian micro scale enterprises registered officially is 622,209 representing 94.4% of the total number of companies registered in the country (Corral *et al.*, 2005). SMEs contribute with 42% to the gross domestic product (GDP) (Su, 2004). Thus, SMEs, including microenterprises, constitute an important portion of the business sector in Peru, not only in terms of numbers but also in terms of employment and contribution to GDP.

Organized in associations, clusters, cooperatives and single companies, SMEs are prominent in several productive activities such as agriculture, agri-industry and agri-exportation, forestry and manufacturing, aquaculture and fishing, fiber, textile and garment, mining, metallurgy and metal working, waste management and water treatment, energy supply, and tourism and craft-making (Su, 2004). According to Corral *et al.* (2005) 12% of the SMEs are dedicated to productive actives, the others are engaged in commercial and services activities.

However, next to their positive economic role, SMEs are also responsible for significant disturbances of nature, environmental degradation and threats to human health (CEPAL, 2006; Pimenova & Van der Vorst, 2004). Environmental pollution related to the increase of productive activities has become evident in Peru and the entire region of Latin America (UNEP, 2003). Air and water pollution are the most important sources of health impacts in Peru (Liebenthal & Salvemini, 2011). Chemical soil pollution is also an important environmental problem in the region related to the growth of agri-SMEs using agrochemicals over the past 30 years. When compared to global consumption, Latin America is an important consumer of fertilisers, representing 9% of the total world consumption and with an annual growth rate of 4%. Agricultural production causes land degradation, increases deforestation, pollution and emissions of greenhouse gases, diminishes biodiversity, and further exploits natural resources. According to UNEP (2003), agrochemical pollution has impacts on soil, water and on the human health of farmers and consumers. Moreover, environmental pollution by SMEs, particularly textile, leather, food and metal transformation industries, is related to low quality employment in terms of productivity, income and working conditions (Minaya, 2007; Zucchetti & Alegre, 2001). Low environmental performance of SMEs impacts on their business performance and competitiveness in a globalized market (Flores, 2004; Su, 2004; TTN-Red Andina, 2005). Thus, moving from polluting SMEs to sustainable SMEs, where environmental rationality goes hand in hand with economic rationality, is an urgent need in Peru and the entire Latin American region.

Because of such environmental challenges, Latin American governments have developed regulations and implemented command-and-control strategies for the industrial sector in order to include environmental considerations as a part of the companies' overall management. Peru has taken a number of initiatives to further integrate the different elements of its environmental management framework, including the establishment of the Structural Framework for Environmental Management of 1993, the National Environmental Management System Law of 2004, the General Environmental Law of 2005 and the creation of the Ministry of Environment (MINAM) in 2008 (Liebenthal & Salvemini, 2011). Nonetheless, the country faces significant challenges in terms of controlling pollution and advancing sectoral environmental management, and most importantly, in addressing environmental health impacts, disaster prevention, and risk mitigation (Liebenthal & Salvemini, 2011). Implementation of regulatory strategies has not been effective in treating environmental pollution problems in the whole business sector. According to IDB (2008), national governments are facing institutional weaknesses, implementation gaps of environmental policies, shortage of financial resources, and operational limitations in law enforcement and reaching SMEs targets. From an institutional viewpoint, there is a lack of collaboration with other actors and resource allocation bears little relationship to sector priorities. For instance, national governments and NGOs collaborate scarcely and their relationship is characterized by tension and power struggle (Bebbington & Thiele, 1993). According to CEPAL (2010) public policies for supporting SMEs in Latin American countries are not effective. Also SMEs need to be part of the agenda of priorities of the Peruvian government, and not only large extractive activities such as mining and oiling exploitation (IDB, 2008).

SMEs are facing several challenges due to low efficacy of national governments. In Peru, SMEs are facing problems such as lack of support, capital, skilled human sources and technology, unfair foreign competition, access to credits and limited access to national and international markets (Korhonen, 2000; Villarán, 2001). For instance, expenditures in research and development (R&D) for industrial environmental upgrading in Peru, including SMEs, over the last decade is estimated at 0.22% of GDP. This percentage is below the average of 0.64% in developing countries and of 2.92% in developed countries (Su, 2004). Until 2004 the investment of Peru in science and technology which is related to the development of national SMEs has been about 0.15% of GDP, which is lower than the regional average of 0.6% (CONCYTEC, 2009). Only 5% of the 60 millions SMEs have access to credit in Peru (Su, 2004). To generate quality employment and sustainable economic growth an investment of US$ 4,000 millions (8% of GDP) in SMEs development is needed (Su,

2004). Moreover, national programmes of support for SMEs do not reach most SMEs (Corral *et al.*, 2005), which makes tackling environmental problems even harder.

The limited success of regulatory strategies in tackling industrial' environmental problems has made that in several countries governmental policy-makers embrace pollution prevention and non-regulatory environmental management concepts, in the hope that they can bring environmental relief by getting to the root of the problems (Brilhante, 2001). In environmental policies worldwide, increasing attention is paid to the role of other societal actors than governments in promoting and enabling conditions for environmental friendly practices for the industrial sector (Brilhante, 2001; Thorpe, 1994; Verheul, 1999). In this new approach of environmental governance, the process of internalizing the environmental dimension in the economic functions of SMEs is not only the role of governments but also of companies themselves and other non-state actors. Based on such an approach, the Peruvian government has encouraged industries to adopt environmental standards, cleaner technologies and energy saving measures in partnership with international cooperation agencies, businesses and local NGOs (MINAM, 2011).

As these developments show, actors such as NGOs, business associations, labour unions, CBOs (community-based organizations) and universities can contribute significantly to creating conditions for sustainable SMEs in developing countries. Particularly NGOs are of special interest because they are organized in platforms and networks of collaboration, from local to global and thus reach beyond the scope of national governments. Additionally, NGOs are diverse in their perspectives and often believed to be not as bureaucratic as national governments; so they can specialize in addressing particular needs of SMEs (Bebbington & Thiele, 1993). Therefore, there is much potential in the strategy of NGOs towards cooperation with and supporting SMEs to move into more sustainable directions. In Peru, several NGOs are very active in promoting sustainable SMEs. A well-known example in which several local NGOs have been supporting a cluster of small enterprises to improve their environmental performance is the 'Villa El Salvador Industrial Park' in Peru (Condor, 2001). Moreover, collaboration efforts towards SMEs in Peru do not only involve local NGOs but also international ones, such as OXFAM, HIVOS, CORDAID and ICCO. Usually, this active role of NGOs in improving environmental performance of SMEs is strongly linked with business strengthening, community development and poverty reduction strategies (Lee, 1998).

As NGOs get more embedded in supporting productive activities of SMEs for local and global markets, they are changing their traditional roles. NGOs are evolving from a social focus to a market focus. Before the 1990s NGO intervention in Peru has been strongly based on philanthropy and charity (Sanborn & Portocarrero, 2008). NGOs were linked strongly to peasant movements and social liberation. This pattern changed in the 1990s, when the neoliberal structural adjustment policies that came into Latin America pushed Peru and other neighbouring countries to open their economies (Jiménez, 2001). Donors often preferred to work with NGOs, claiming that government counterparts were inefficient. As a consequence a strong shift took place in the way NGOs and other international cooperation agencies deliver their developmental mandates. While some NGOs kept to the ideology and motivation of the 1960s and 1970s, other changed course; also new and more opportunistic types of NGOs emerged following these changes (Bebbington & Thiele, 1993).

These developments in Latin America mirrored similar developments in other parts of the world, where NGOs emerged to perform a range of crucial roles in greening SMEs. NGOs acted as intermediary organizations and 'green' stimulators by supporting SMEs with technical support for sustainable production and eco-efficiency, guidelines for good house keeping, codes of conduct, higher value products, green labelling, 'green' (micro-) credits and employee training. NGOs also facilitated links of SMEs with business partners, governmental authorities, and development agencies, and promoted institutional conditions under which SMEs can improve their environmental performance, innovation, productivity and competitiveness (Bianchi & Noci, 1998; Brio & Junquera, 2003; Frijns & Van Vliet, 1999; Gombault & Versteege, 1999; Hyman & Dearden, 1998; Kaimowitz, 1993; Korten, 1987; Mol, 2000; Vonortas, 2002). Moreover, NGOs also gave a voice to their SME target groups in the media, thereby influencing citizen and consumer opinions (Landim, 1987).

On a more theoretical level, such developments imply that the market has become a new battle-field for NGOs in their efforts to support sustainable development of SMEs. Focusing on social and economic actors beyond the government means for NGOs to explore new ways to advance the environmental governance of production in a globalizing context (Buttel, 2001; Mol, 2000). NGOs are becoming central actors in social and economic development, overtaking tasks of national and local governments and engaging with economic actors, such as local enterprises (Mol, 1995). This increasing role of NGOs in engaging with productive activities and market processes challenges conventional theories of governance which centred on the national government, as well as conventional theories of NGOs which centred on conflict and protest strategies.

However, these changes mean also uncertainties for NGOs. According to Bebbington and Thiele (1993) the changes of NGOs in Latin America have danger to undermine their legitimacy and even the very credibility of the label NGO. Many NGOs are driven today less by ideals and more by income-enhancing concerns of a professional middle class staff. New types of NGOs with opportunistic features would be more than willing to fill the gaps left by a receding state, as long as they are paid to do so. Bebbington and Thiele (1993) highlight that the risk for NGOs is that they will end up performing new instrumentalist roles, stimulated by donors and governments, rather than a genuine role as agents of democratization.

From this introductory overview, it becomes clear that there is a need to further investigate the actual and potential roles of NGOs in the environmental transformation of SMEs. The potential of NGOs in the environmental governance of SMEs is not fully realised. On the one hand a complex set of factors is limiting the innovative power of SMEs, while on the other hand NGOs have to reinvent their conventional roles in a global context of shifting agendas of business and government. More knowledge is needed on the actual roles, impacts and degrees of success of NGOs in promoting SME environmental improvement, especially in developing countries (Bianchi & Noci, 1998; Condor, 2001). As Vakil, (1997) argues, answers to these concerns may well lead to a distinctive body of theory on organization strategies pertaining to NGOs as special types of organizations.

To shed more light on actual and potential roles of NGOs in this changing societal context, this research explores the changing roles of NGOs in promoting sustainability of SMEs in Peru. The research is based on two theories that deal with non-state actors: ecological modernization theory (e.g. Mol, 2000) and network society theory (e.g. Castells, 2000). Perspectives of networks and discourses are used as analytical tools, with which these theories are applied to understand the changing role of NGOs towards greening SME production. The research focuses on three domains of environmental reform of SMEs: organic production, sustainable production and business social responsibility. All three domains represent promising approaches to bring environmental considerations to bear on production and markets. While the approaches share commonalities in engaging a wide range of local and global actors involved in the economic and social development of SMEs in developing countries, each approach has its own claims in bringing environment to production and markets. Indeed, each approach gives a different emphasis to managerial and technological tools to promote sustainable business. For instance, business social responsibility focuses on codes of conduct, guidelines, indicators of sustainability and ISO 14001. Sustainable production focuses on good house-keeping, eco-efficiency and energy saving, among others. Organic production concentrates on organic products and certification and labelling. Moreover, business social responsibility and sustainable production are relating larger business and multinationals to local exporting SMEs. While business social responsibility and sustainable production prioritize SMEs that are keen to reach national and international markets, organic production focuses on SMEs oriented to local and international markers.

Based on the aforementioned theories, perspectives and domains, the research analyzes in depth the changes in networks and discourses of NGOs that support SMEs are undergoing. Central to our analysis are (1) the mechanisms by which networks of NGOs are collaborating to support the sustainability of SMEs in Peru, and (2), the discourses that NGOs utilize to justify and shape their approach to promote sustainability in SMEs. This analysis will help to provide theoretical and empirical knowledge to better understand and explain the actual, new and potential roles of NGOs and the social-political relations, dynamics and inter-organizational relations between NGOs, SMEs and other societal actors in the greening of SMEs. Also, this study analyzes the implications of changes in the roles of NGOs for the environmental governance of SMEs in Peru and Latin America.

1.2 Research aim and research questions

The central aim of this research is to provide a better understanding of the changing roles of NGOs in promoting sustainability of SMEs in Peru. The perspectives of networks and discourses will be used for this.

The main research questions in this thesis are:

a. What are the networks of NGOs promoting sustainability of SMEs involved in the domains of organic production, business social responsibility and sustainable production in Peru, and what are the main changes in time in these networks?

b. What are the main discourses fostering sustainability that prevail and are articulated in these networks of NGOs and what are the main changes in time in these discourses?

c. How to understand and assess the actual, new and potential roles of NGOs in promoting sustainability of SMEs in terms of network society theory and ecological modernization theory?

1.3 Scope of research

NGOs, as a concept used in this research, includes a broad and heterogeneous set of non-profit organizations ranging from informal social movement organizations to highly professionalized ones, from local (national) to international organizations, and from northern to southern NGOs. In this study the universe of NGOs is narrowed to NGOs operating in Peru, organized in networks at national, regional and global scale, and aiming to support local SMEs in their aim for sustainability. Those NGOs provide support (a) to medium and small scale producers and producer associations to bring organic products to local and global markets, (b) to urban and rural small scale enterprises to adopt cleaner production and appropriate technologies, and/or (c) to SMEs to upgrade social and environmental standards within value chains involving large companies.

For the purpose of this research, SMEs are defined by the number of employees. To be considered a medium-sized enterprise the number of workers employed are more than 50 and less than 100 workers; a small sized enterprise has more than 10 and less than 50 workers (Teszler, 1993; Uribe-Echevarria, 1991). This research also includes micro-enterprises with less than 10 workers. Thus defined, SMEs include among others rural small scale enterprises, urban small scale enterprises, small scale producers, small family holders, small and medium agri-industries, garment workshops and craft-makers. This research particularly focus on SMEs that are internalizing sustainability by improving their social and environmental performance and are supplying products for local and international markets in collaboration with NGOs. In specifying the scope of the research, it is worth noting that in developing countries micro scale enterprises are socio-economically related to small-sized enterprises and small-sized enterprises are socio-economically related to medium-sized enterprises. Furthermore, this last group of enterprises are the minority of SMEs in number, but not necessarily in contribution to GDP. Because of this particularity, research and public policies focus more on the 'cluster' of small and micro scale enterprises (the so called 'MYPE' in Latin America, a Spanish acronym) than on the entire SMEs. This is also the case in this research. So when I use the term 'SMEs' in this study, I most often refer to small-sized and micro scale enterprises, though not excluding medium-sized companies.

The research is not limited to a certain geographical area within the country, neither to a particular production domain, since NGOs carry out projects throughout the country and support several types of SMEs. Some SMEs are concentrated in the main cities of Peru such as Trujillo, Arequipa and Lima, while other SMEs, such as organic food producers, are spread all over the country. In any case, SMEs under this research have collaboration ties with the NGOs to be studied.

Networks, in this thesis, are understood as composed of collaborative relationships and exchange of resources between actors promoting sustainability of SMEs. Usually, the actors involved in these networks of sustainability of SMEs are NGOs, SMEs, international cooperation agencies, large companies and governmental agencies. Networks in this study have often a 'hybrid' character as NGOs are usually considered part of civil society and companies are usually considered part of the economic domain. The study of NGO networks starts from the country level, but also involves local networks and regional and global networks. The NGO networks are often organized around platforms at national, regional and global scales. Both key platforms and key individual organizations are defined as central nodes in these networks.

Discourses, in this thesis, are understood as storylines articulating the promotion of sustainability of SMEs, with specific perspectives on markets, environment and production. Discourses are a coherent set of views, changing in time, and constraining and enabling the adoption of sustainable practices of SMEs. Discourses of sustainability of SMEs can be and are shaped by local and global agents; they can depart from conventional development discourses but can also extend and enrich them. Discourses are studied to uncover the rationality behind the support strategies of NGOs towards sustainability of SMEs in the context of markets.

1.4 Overview of the thesis

The thesis is organized in nine chapters. Chapter 2 provides an overview of the theories framing this study: ecological modernization theory and the network society theory. Chapter 3 presents the methodology and describes in detail the concepts of social network analysis and discourse analysis and the key variables to be analyzed for mapping the networks and discourses of sustainability of SMEs. The chapter also includes the case selection and the data collection methods.

Chapter 4 provides an overview of the domains of organic production, business social responsibility and sustainable production, with special attention to the international level. Subsequently, the three case studies are presented. The case of organic production is presented in Chapter 5, the case of sustainable production is presented in Chapter 6 and the case of business social responsibility is presented in Chapter 7. Each case study maps the networks and discourses with a focus at the national level, but involving local and global linkages.

In Chapter 8 the discourses and the networks promoting sustainability of SMEs in Peru are analyzed comparatively. Based on this comparative analysis, the roles and perspectives of NGOs are investigated in relation to the Peruvian state and to international cooperation agencies. Finally, in Chapter 9 the research questions are answered by drawing main conclusions in terms of shifts in networks and discourses. In addition, a theoretical reflection on the outcomes of the research is presented. The chapter also includes recommendations for policy and research on the role of NGOs in the environmental governance of SMEs worldwide.

Chapter 2.
Analysing changing NGOs-SMEs interactions: theories, concepts and a framework

2.1 Introduction

In this study ecological modernization theory and network society theory are applied as research frameworks to shed light on the changing roles of NGOs promoting sustainability of SMEs in Peru.

Ecological modernization theory helps to understand changes in discourses, with a focus on three main issues: the institutional transformation of NGOs as agents improving environmental conditions of SMEs, the reinforcing position of environmental rationality in the development of SMEs and the access and development of sustainable markets by SMEs. It helps to understand under what conditions NGOs are effective in catalyzing environmental reform of SMEs. As NGOs are organized in networks of global reach, network society theory offers insights in contemporary changes in governance arrangements, and in cooperation and power relationships that affect the networks of NGOs.

The chapter will first give an overview of these theories, and then discus their application to this research.

2.2 Ecological modernization theory

Ecological modernization theory is one of the major sociological theories aimed at understanding how modern societies are dealing with environmental reform. Environmental reform encompasses the notion that modern industrial societies can solve environmental problems through intensified development of innovative industrial technology, through modification of processes of production and consumption according to ecological criteria and the application of ecological efficiencies measures, and through green marketing and other strategic environmental management practices (Buttel, 2001; Mol, 1995).

During the early development of ecological modernization theory in the 1980s the emphasis was on the role of technological innovation, especially in the sphere of industrial production. From the late 1980s onward, more attention was given to the institutional and cultural dynamics of ecological modernisation. The emphasis was still very much on national studies regarding West-European countries. From the mid 1990s onward, increasing attention has been paid on the one hand to the global dynamics of ecological modernisation, and on the other hand to national studies in developing countries, Central and East European nation-states, and the USA (Huber, 2000; Mol, 2000, 2001; Spaargaren & Mol, 1992).

Below, three key concepts of ecological modernization theory are presented in more depth, because they are of particular relevance to this study: institutional transformation, environmental rationality and market dynamics.

2.2.1 Institutional transformation

Ecological modernisation theory recognizes the structural character of environmental problems but nonetheless assumes that existing political, economic, and social institutions can internalize environmental protection. To restructure modern societies on an ecological basis, the institutions of modern society, such as markets, the state and science and technology, need to be radically transformed (Mol, 2000). According to ecological modernization theory institutional transformation means redefinition of roles of the state *vis-à-vis* market actors and NGOs (Mol, 1995, 2000; Seippel, 2000). This transformation implies furthering ecology into the process of modernization, according to some in a so-called process of hyper-or-superindustrialization (Mol, 1995). It also implies a transition of modernity to a reflexive modernity phase (Spaargaren, 2000) that will be radically different from the simple modernity phase which prevailed during most of the twentieth century. Thus, ecological modernization theory incorporates the idea of reflexive modernization in dealing with the ecological crisis (Mol, 1995; Mol & Spaargaren, 2000).

An important aspect of institutional transformation is the increasing role of non-state actors in environmental policy and the replacement of the old hierarchic state model with new governance styles (Mol & Spaargaren, 2000). Under these new governance styles, the state, businesses and NGOs are identified as companions in environmental struggles rather than enemies (Mol, 2000). Ecological modernization theorists observe that the current environmental movement is shifting from radical opposition to more cooperative stances toward industry, encouraging environment-oriented business practices (Buttle, 2001; Grohmann, 1997; Hallstrom, 2004; Mol, 2000). As a result this change in ideology and strategy seems to provide NGOs with better access to both the general public and the core of policy communities in the environmental field (Mol, 2000).

2.2.2 Environmental rationality

Environmental rationality means focusing on re-directing economic processes according to ecological criteria and towards ecological goals. To institutionalize environmental rationality in production and consumption processes, and thus to redirect economic practices into more ecologically sound ones, the ecological sphere, characterized by ecological rationality, has to be emancipated from the economic dimension of modernity (Mol, 1995; Mol & Spaargaren, 2000; Spaargaren, 2000). The emancipation of ecological rationality is not perceived as a process towards the dominance of ecological rationality or the ecological sphere over or instead of the economic rationality and sphere. Each one has its own goals, realm and legitimacy (Mol, 1995). The social practices of production and consumption should be designed and evaluated according to both environmental and economic rationalities (Mol, 1995). Furthermore, environmental rationality is no longer automatically linked to, nor can it be reduced to, more traditional political, economic, social or other rationalities, or to ideologies such as socialism, conservatism and liberalism (Mol, 2000).

There are two main mechanisms in the re-structuring of economic practices according to principles of ecological rationality. These mechanisms are called ecologizing of economy and economizing of ecology. Ecologizing of economy refers to the development of new and more environment-friendly technologies and processes. It includes several strategies, such as more efficient material and energy use, closing of substance cycles and the monitoring of material and energy flows. It also may imply substitution of ecologically maladjusted technical systems and economic sectors. Economizing of ecology refers to placing an economic value on nature and environmental pollution. It include several strategies, such as eco-taxes, environmental liability, insurance for environmental care, technological transformation and change such as cleaner production, ecological sound products on the market, environment as factor in economic competition, environmental audits as precondition for commercial loans and economic investments, and environmental accounting and bookkeeping (Mol, 1995).

The restructuring process should result in the institutionalization of ecology in the social practices and institutions of production and consumption. Through these processes, as ecological modernization posits, it is possible to combine economic prosperity and environmental protection. Environmental protection is no longer seen as a burden upon the economy, but rather as a potential source of future growth (Langhelle, 2000). Thus, a sound environmental performance is a precondition for sound economic performance and, *vice versa*, that taking care of the environment could mean profitable business (Seippel, 2000).

2.2.3 Market dynamics

Market is a key concept in ecological modernization theory. In the process of ecological modernization, liberal market capitalism is redirected and transformed in such a way that it less and less obstructs, and increasingly contributes to, the preservation of society´s sustenance base (Mol & Spaargaren, 2000). Market dynamics are seen as a social carrier of ecological restructuring and reform (Mol, 2000).

The theory highlights the new role of the market in triggering environmental protection (Mol & Spaargaren, 2000). Not only do the emerging environmental interests open up new markets and create new demands, it also stimulates innovation in methods of production, industrial organization and market supply. Rather than a threat for the system, markets become a vehicle for its innovation (Hajer, 1995). According to Mol and Spaargaren (2000), market forces, innovative entrepreneurs, consumers, insurance companies, creditors and commissioners, among others, will emerge as new important forces in and drivers and social carriers of the ongoing process of socio-ecological transformation. Examples of profitable market activities are the expanding eco-innovations in the chemical industry, the development of environmental technologies and the provision of ecological advice (Mol, 1995).

According to ecological modernization theory, the market is considered to be a more efficient and effective mechanism for coordinating the tackling of environmental problems than the state (Mol, 1995). Being rooted in the principles of a preventive environmental policy and a social market economy (Andersen & Massa, 2000), ecological modernization theory argues for a transfer of certain responsibilities, incentives and tasks from the state to the market. The state provides the necessary conditions and stimulates social self-regulation, either via economic mechanisms or via the public sphere of citizen groups, environmental NGOs and consumer organizations. The shifting traditional role of state worldwide is one of the factors contributing to the shifting roles of NGOs.

2.3 Network society theory

Network society theory argues that the current society constitutes a new type of social structure characterized by two emerging social forms of time and space. These are timeless time and the space of flows. Timeless time is defined by the use of new information technologies (Castells, 2000). The space of flows organizes the simultaneity of social practices at a distance, by means of telecommunications and information technologies. The space of flows is not placeless. It is made of nodes and networks which are connected to the space of places. The space of places is the material support of time-sharing social practices where meaning, function and locality are closely interrelated. It privileges social interaction and institutional organization on the basis of physical contiguity (Castells, 2004). What is distinctive of the new social structure, the network society, is that most processes, including power concentration, wealth creation and distribution, and information exchange are organized in the space of flows. Nevertheless, most human experience, and meaning, is still locally based. According to network society theory, there is a fundamental opposition emerging in the network society between two logics of space, that of the space of flows and that of the space of places. The disjunction between the two spatial logics is a fundamental mechanism of struggle in our societies, because the core economic, symbolic, and political processes shift away from the realm where social meaning can be constructed (Castells, 1997). Thus, dominant functions in current society are organized in networks pertaining to a space of flows that links the networks around the world, while fragmented subordinate functions and people are located in the multiple spaces of places. According to Castells (2004), the main option that the local has to avoid being overwhelmed by the global is to become a node in alternative global networks.

In network society theory, networks are defined as a set of interconnected nodes and open structures able to expand without limits, integrating new nodes as long as they are able to communicate within the network. Thus, a network-based social structure is a highly dynamic, open system, susceptible to innovating without threatening its balance. Networks can be seen as an organism where all the individual units (nodes) of the network cooperate to achieve their goal (Castells, 2000). Flows are the streams of information and other resources between nodes circulating through the channels of connection between nodes (Castells, 2004). Networks are very old forms of social organization but only with the recent development have they become the dominant form of social organization. On the one hand, networks are flexible and adaptable forms of organization, able to evolve with their environment and with the evolution of the nodes that compose networks. On the other hand, networks have considerable difficulty in co-ordinating functions, in focusing resources on specific goals, and in managing the complexity of a given task beyond a certain size of the network (Castells, 2000).

Below, three key concepts of network society theory are presented in more depth: governance arrangements, cooperation and power relationships.

2.3.1 Network governance arrangements

According to network society theory, economies throughout the world have progressively become globally interdependent, introducing a new form of relationship between economy, the state, and society, in a system of variable organizational structures and strategies. Network society theory emphasises organizational transformation and the emergence of a globally interdependent social structure. What matters in this new social arrangement is the networking capacity of institutions and organizations, both locally and globally (Castells, 2004).

Particularly, the relationships of production and consumption are structured under new governance arrangements (Castells, 2000). Thus, the diffusion of networking logic in processes of production substantially modifies their operation and outcomes. Production is defined in network society theory as technological arrangements. It means a set of tools, rules, and procedures through which scientific knowledge is applied to a given task in a reproducible manner (Castells, 2000). Production is the action of humankind on nature, to appropriate it and transform it for its benefit by obtaining a product, consuming part of it, and accumulating surplus for investment, according to socially decided goals (Castells, 2000). Currently activities of production, consumption, as well as their components (e.g. capital, raw materials, management, information, technology, markets) are organized on a global scale, either directly or through a network of linkages between economic agents. In fact, capitalism itself has undergone a process of profound restructuring, characterized by greater flexibility in management, decentralization and networking of organizations in their relations. In this new context, despite the fact that the network society is the dominant new paradigm, industrialism does not disappear. Indeed, industry is seen as fundamental component of the new paradigm (Castells, 2004).

As a result of such re-arrangements, during the last years environmental networks organized in regional and global platforms have been emerging as part of the new societal spaces. These global environmental movement networks are diverse in their composition, and vary widely in their expressions from country to country and between cultures. Castells (1997) argues that there is a relationship between the rise of environmental global networks and the fundamental dimensions of the network society. Due to the pervasive profile of the space of flows, most environmental NGOs have become largely institutionalized, that is, they have accepted the need to act in the framework of existing institutions, and within the rules of the global market economy. The environmental movement is at the same time localist and globalist: globalist in the management of time and localist in the defence of place. Much of the success of the environmental networks comes from the fact that they have been able to best adapt to the conditions of communications and mobilization in the new technological paradigm (Castells, 1997).

Business has been influenced by changes in the environmental movement, and has tried to adapt their process and their products to new legislation, new tastes, and new values, while trying to make profit out of it at the same time. An example of adaptation of business to the new conditions is the endorsement by companies of emerging approaches promoting sustainability, such as business social responsibility and cleaner production. Furthermore, the increasing organization of business in global value chains such as food and garment is contributing to decentralize environmental transgression to SME suppliers spread in several countries all over the world. As Castells (1997) states, the growth of production and the flow of supply of SMEs towards the global markets have implications on the distribution of the environmental burden. Therefore, to influence properly the global business networks, environmental NGOs have to decentralize their action from the centers of global consumption to the 'tails' of global production. This means NGOs (aim to) influence not only the business headquarters but also their SME suppliers. To do so, environmental NGOs, like business, have to be organized in networks of global reach. Furthermore, to be effective in tackling environmental burdens of global supply chains, global networks of NGOs need to include SMEs in the decision making. Therefore, improving the environmental conditions of SMEs need a close coordination of local NGOs, international NGOs, SMEs and larger companies involved in value chains.

2.3.2 Network cooperation

According to Castells (2004), cooperation in networks is based on the ability to communicate. This ability depends on the existence of codes of translation and inter-operability between networks (protocols of communication), and the access to connection points (switches) (Castells, 2004).

Cooperation has become a key issue in the network society. For instance, economic networks establish alliances, agreements and joint ventures in order to have access to profitable markets and to be able to compete (Castells, 2000). Major companies work in a strategy of changing alliances and partnerships, specific to a given product, process, time and space. Similarly, SMEs are connected in networks. Such networks are often established for the purpose of specific business projects, e.g. a particular production process, and disappear or are modified into another network as soon as the project is finished (Castells, 2000). The unity of the production process is not the firm but the business project. The firm become a node that is part of global economic networks and flows (Castells, 2000).

Network society theory acknowledges cooperation between environmental networks and business networks. Over the last few decades, both networks have been working in coordination and have put their perspectives and needs in a common agenda. Environmental networks focus on making environmental legislation and governance by lobbying the state and collaborating with companies (Castells, 1997). None of the actors involved in the networks has the capacity on its own to bring about the changes necessary for environmental reform. Therefore, in environmental reform business and NGOs have starting to move to strategies of cooperation and trust rather than confrontation (Brand & De Bruijn, 1999).

2.3.3 Network power relationships

Power is understood in this research in terms of flows of resources, connections and values. Particularly, resources are exchanged through the networks as flows of information, funding and technical expertise. The capacity to be resourceful and allocate resources puts actors in position of power. What resource is powerful and powerless is defined by key actors. Key actors are the power holders in the networks, performing as network nodes. As influential actors, they are in best position to define the aim and the configuration of the networks. If a key actor in the network ceases to perform a useful function, it is phased out from the network, and the network rearranges itself. Some nodes are more important than others, but they all need each other as long as they are within the network (Castells, 2000). Therefore, networks work on a binary logic: inclusion/ exclusion. All there is in the network is useful and necessary for the existence of the network. What is not in, is ignored or eliminated (Castells, 2000). Nodes increase their importance for the network by absorbing more relevant information, and processing it more efficiently. The relative importance of a node does not stem from its specific features but from its ability to contribute to the network goals (Castells, 2004).

Networks have no centre and actors share decision-making (Castells, 2000). While some actors have a greater degree of influence than others, there is never an absolute power or zero degree of influence of one actor over another (Castells, 2009). Networks make it materially impossible to exercise hierarchical power without processing instructions in the network, according to the network´s rules. While hierarchical power implies a chain of command among actors, democratic power implies a shared decision making among actors. Thus, in the network society, power is redefined, but it does not vanish. Nor does social struggle. Each network defines its own power system depending on its programmed goals (Castells, 2004).

The ability of exercising control over others in the network depends on two basic mechanisms: the ability to program/reprogram the networks; and the ability to connect different networks to ensure their cooperation by sharing common goals and increasing resources. Castells (2004) calls the holders of the first power position 'programmers' and the holders of the second power position 'switchers'. Network nodes are the key actors and because of their position in the social structure, they exercise power in the network society (Castells, 2004).

The power relationships in networks are also related to the dynamics of domination and of resistance to domination. Dominant functions are no longer organized in geographical spaces but rather in the space of information networks (Castells, 2000). This brings togetherness and separation of networks at the same time. According to Castells (2000), the dialectical opposition of space of flows and the space of places provokes simultaneous processes of globalization and localization. In the struggle for domination, environmental networks emerge as challengers of the dominant structure of the network society, which is made up of the global financial market networks and the network enterprises.

Power relationship is particularly central in the social structure of production and consumption (Castells, 2000). Political institutions are not the main site of power any longer. The more decisive power is the power of information and communication flows, and cultural codes embedded in networks (Castells, 2000). Actors, rooted in the social structure of the network society, influence the relationships of production and consumption by enacting, reproducing or transforming it. This process, naturally, involves actors in sharp conflicts and strong disagreements with each other (Castells, 2000). Political institutions embodied in the national government have partly retreated from governing production and consumption structures in favour of global (financial) market networks. This increasing loss of power of governmental structures sets the stage for a more dynamic role of the networks of NGOs in influencing production and consumption structures. Particularly, NGOs are challenged to stand for the social and environmental concerns caused by such structures. This can be done not only by pressuring global market networks but also by making alliances with companies endorsing sustainability and 'green' products and consumption. Thus, the power struggle is not only between the NGO networks and business networks but also between 'frontrunner' companies and 'laggard' companies, and between NGOs opposing market dynamics and NGOs endorsing it.

2.4 Application of ecological modernization theory and network society theory to the research

Before explaining the applications of ecological modernization theory and network society theory to the research, the main features of both theories are presented in the Table 2.1.

The application of ecological modernization theory and network society theory to the research is organized in three themes, each of which is particularly linked to specific key concepts that were elaborated for each theory – though all themes also build on the wider argument developed above. The first theme deals with governance arrangements for environmental reform of SMEs, the second with cooperation of environmental NGOs and SMEs, and the third with shifts from hierarchical power to more democratic ways of power.

2.4.1 Governance arrangements for environmental reform of SMEs: larger, diversified and flexible roles for NGOs

This theme is particularly linked to the aforementioned key concepts of 'institutional transformation' and 'network governance arrangements'. Building on these notions, the following arguments can be made.

First, the institutional transformation of NGOs contributes to increasing their potential roles. The main institutional changes that NGOs are undergoing are: change in the relationships with business from confrontation to cooperation; the acceptance of the market as playground for development efforts towards SMEs; and the move from isolated work to more coordinated work with other actors. In supporting SMEs, NGOs increasingly work with several actors including

Table 2.1. Main features of ecological modernization theory and the network society theory.

Dimension	Ecological modernization	Network society
Core	environmental reform	global networks
Emphasis	cooperation	struggle, inclusion/exclusion
Focus	institutional transformation	institutional domination and resistance
Economic system	capitalistic	capitalistic
Social structure	modernity	post-modernity
Material structure	industry	information technology
Claim	reform of capitalism	new era
Environment	independent rationality	part of wider social sphere
Environmental NGOs	central in reform of modernity	alternative networks
Power	business, NGOs	network enterprise, global financial network
Scope	industrialized and industrializing countries	global scope
Subscribe	reflexive modernity	social justice

multilateral cooperation agencies, international financing organizations, grant-making foundations, larger companies and governmental agencies. Those changes have opened new fields of action for NGOs, especially fulfilling the needs of SMEs at the market.

Second, the diversification of roles of NGOs is expected to answer the heterogeneity in characteristics and types of SMEs. The diversification of roles has to do with two processes that SMEs are undergoing: the increase in linking and embedding of SMEs in global market networks and the growth of SMEs anchored in the geographical space. NGOs can support global market oriented SMEs, the so-called 'exporting' SMEs, by performing as market facilitators and market intermediaries. NGOs might engage with business associations to lead self-regulatory initiatives, and with citizen associations to empower consumers. As SMEs' production becomes globally oriented, NGOs increasingly target them to support the reduction of their environmental impact, the meeting of international standards and facilitating linkages between local SMEs and global buyers. On the other hand, NGOs can also support place-bound SMEs, the so called 'surviving' SMEs, by linking them with the space of flows. Actually, networks of NGOs promoting sustainability of SMEs are part of what Castells calls the space of flows, but anchored in the space of places. To be effective in making the links, NGOs with local roots are needed. While local NGOs are rooted in the geographical space, international NGOs are mainly part of the space of flows as they intend to influence SMEs located in far away places. In supporting surviving SMEs, NGOs might provide assistance in the development of low-cost production methods and technologies, the efficient use of natural resources and the identification of niche sustainable markets for their products. The fostering of environmental reform for both the exporting SMEs and surviving SMEs, has to include, as Mol (1995) claims, ways to de-link SMEs' economic growth with their environmental degradation.

Third, NGOs have flexible roles in supporting SMEs by combining cooperation and pressure strategies. Environmental NGOs usually have emphasized pressure strategies against business and the state. As the market has become central in economic relationships, traditional roles of NGOs have changed and new roles have also emerged. Open confrontation has been replaced by monitoring, influencing and advocacy. New roles based on cooperation, such as alliances, partnerships and join ventures, have emerged. Therefore, environmental NGOs, as part of the environmental movement, have obtained flexible roles. On the one hand, they challenge the power of global markets, but on the other hand they collaborate, for instance, by helping SMEs to access these markets.

Fourth, the potential of SMEs to adopt sustainable practices has to be realized in the design of reflexive governmental arrangements. Exporting SMEs are usually seen as forefront SMEs in adopting sustainable practices. According to Castells (2000), most of the initiatives among SMEs on environmental management strategies, for instance, on eco-labelling schemes, have been reported to come from the exporting SMEs. However, SMEs are seen as a business sector facing more constrains than larger companies. While in all major economies larger companies are at the centre of the structure of economic power, SMEs are seen as marginal economic actors with limited competitiveness. Nonetheless, SMEs have showed to be resilient agents of innovations and job generation. SMEs appear to be forms of organization well adapted to the flexible production system of the global economy. Actually, larger companies increasingly are subcontracting SMEs, whose vitality and flexibility allows gains in productivity and efficiency (Castells, 2000). On the other hand, products exported from developing countries are made not only in formal enterprises but also in the informal sector with small, possibly household-level, units. Certainly, as Hale (1996) argues, engaging SMEs in environmental reform would be a challenge, especially for large number of 'surviving' SMEs in developing countries, which belong to the informal sector. However, surviving SMEs have also potentialities to be engaged in environmental reform. The principle to address environmental reform in surviving SMEs is that environmental protection and the conservations of natural resources are the bases not only to survive in the market but to become competitive. As surviving SMEs have experience in doing business and know well their local environment, they can easily implement changes and innovations in their production processes and technology appropriate to their needs in line with environmental goals. As a result, surviving SMEs might orient their products from low lucrative markets to profitable sustainable niche markets. Additionally, next to technological considerations, social and cultural aspects have to be considered to address properly environmental improvements in this type of SMEs. This analysis of the potential of SMEs to endorse environmental reform shows the heterogeneity of SMEs in developing countries and, therefore, no universal governance arrangement can optimally address it. A feasible model of environmental governance has to encourage the increase the production scale of SMEs, the building up of SME structures for reform, and the design of strategies of centralization and decentralization of SMEs production. Therefore, cooperation of SMEs and NGOs in working towards environmental reform implies the development of 'reflexive' environmental governance arrangements tailored to the heterogeneity of SMEs.

2.4.2 Cooperation of environmental NGOs and SMEs in developing and accessing sustainable markets

While the previous theme mainly addressed the governance context of SMEs-NGOs cooperation, we can also apply the theories in formulating expectations regarding the nature of that cooperation, building on the key concepts 'network cooperation', 'ecological rationalization' and 'market dynamics'.

First, environmental NGOs and SMEs will often need each other to achieve their goals. While SMEs provide legitimacy to environmental NGOs, environmental NGOs help SMEs to improve their social and environmental performance. NGOs can cooperate with businesses, including SMEs, to face the heightened social, legal and global market pressures to operate more sustainably. NGOs are good in generating new ideas, mobilizing consumers and organizing public support or disapproval (Mol, 1995). Furthermore, NGOs can be allies of SMEs by launching credible environmental initiatives through 'green alliances'. Green alliances are defined as collaborative partnerships between environmental NGOs and businesses that pursue mutually beneficial ecological goals (Arts, 2002; Brand & De Bruijn, 1999; Stafford, Polonsky & Hartman, 2000). The shared goals of environmental NGOs and sustainable SMEs are to make business sustainable and expand sustainable production and sustainable markets. While not all environmental initiatives lead to competitive gains, a confluence of ecological, social and market objectives is possible from green alliances because they facilitate opportunities for stakeholders to specify problems, discuss needs, establish common ground and implement environmental programs that address the multiple needs of affected parties.

Second, collaborative alliances between NGOs and SMEs need to be oriented – at least to some extend – to developing sustainable local markets and accessing sustainable national and global markets. During the last decade, a set of interrelated issues regarding globalization and trade has increasingly led NGOs into unprecedented alliances with market actors. According to Buttel (2001) environmental NGOs can be most effective if they engage in collaborative relationships with business whose actions have an impact on the environment. In the same vein, Biekart (2005) argues that NGOs might contribute to make SMEs' production and products more environmental friendly by establishing collaborative alliances between market actors, the state and other civil society actors, both in the global north and the global south. NGOs and SMEs might work together in developing sustainable local markets, such as organic farmers markets. NGOs and SMEs might also work in collaboration with larger buyers to access sustainable national and global markets; for instance, supermarkets and malls. The development of sustainable markets for products coming from, for instance, ecological industrial parks and organic farmers markets is important to ensure the creation and nesting of sustainable SMEs. Similarly, the access of environmental friendly SMEs to national and global markets, such as supermarkets, governmental procurement and global 'green' buyers, is important to ensure the expansion of sustainable SMEs. In the former scheme, NGOs might support SMEs by providing knowledge, technology and economic incentives. In the latter scheme NGOs can support SMEs by being engaged with other actors such as larger companies. Particularly in this latter scheme local NGOs, on the one hand, have to simultaneously operate at local and national levels with target groups, including individual SMEs and SMEs associations, while, on the other hand, they need to interact with larger companies and international NGOs at the international level. Moreover, traditional cooperation strategies of SMEs focus on the supply-side, for instance provision of training, credit facilities and infrastructure, and this effectively reaches only a small number of SMEs. According to Frijns and Van Vliet (1999) a supportive strategy for SMEs to access sustainable markets could best focus on the creation of favourable conditions on the supply-side and on the demand-side of the market. Cooperation of NGOs and SMEs might contribute not only to take advantage of such favourable conditions, but also to create them.

Third, the cooperation of NGOs and SMEs is expected to be organized on several levels of network aggregation, in line with a tendency of NGOs and SMEs themselves to be organized on multiple levels. NGOs are often clustered in 'umbrella' (membership) NGOs. The umbrella NGOs are organized in platforms at national, regional and global scope. SMEs are also increasingly organized in clusters of SMEs, SME associations and cooperatives, most of them of national reach. Particularly the cooperation between SMEs is becoming an important feature of economic relations in developing countries (Frijns & Van Vliet, 1999). During the last decades many SMEs have also taken the initiative in establishing networking relationships with several larger companies and/or with other SMEs, finding market niches and cooperative ventures. SMEs have the ability to link up in networks among themselves and with larger companies to be able to compete in the global markets. Also, the cooperation benefits with larger companies within industrial networks can provide SMEs with the needed resources in new areas of business strategy, such as environmental management (Korhonen, 2000). So, NGOs and SMEs are familiar to the network dynamics, though operating usually in separated networks. This acknowledgement of network dynamics will be valuable to start cooperation between their networks.

Fourth, cooperation between NGOs and SMEs might be affected by self-interests and struggles for power positions. Collaboration represents also challenges for NGOs and SMEs. Indeed, the networks are a field of cooperation and struggle among their constituencies. NGOs and SMEs might have different motivations to get engaged in collaboration schemes and different perspectives of sustainability. While NGOs main aim is to promote development, SMEs main aim is to produce products or deliver services at the market. Both aims can complement but they can also clash. The clash has to do with the establishment of priorities, the discourses endorsed and the struggle for power position. Indeed, authors such as Stafford, Polonsky and Hartman (2000), Rossi, Brown and Baas (2000) and Arts (2002) highlight constrains and complexities related to cooperation. This fourth point is also related to the next theme.

2.4.3 Shift of power from hierarchical to more democratic ways of power enables SMEs to develop and access sustainable markets

The final theme to be mentioned is the power relationships that govern the mutual roles of NGOs and SMEs. Here we build on the key concepts of 'network power relationships' and 'market dynamics', among others.

First, power is re-defined in the networks as they get mature. As was described earlier, in a network society hierarchical power among actors will shift to more shared ways to make decisions when a network gets mature. Hierarchical power and democratic power are both present in the networks promoting sustainable SMEs. NGOs and international cooperation agencies used to control power and deliver development towards SMEs based on a top-down approach, while SME used to be passive beneficiaries. In line with Castells, we might expect that networks are undergoing a re-definition of power towards more democratic power among and between NGOs and SMEs. As long as SMEs move from beneficiaries to legitimate actors they will increasingly challenge the power of traditional power holders. In this new context, local and international NGOs have to share power with SMEs and SMEs will gain power *vis-à-vis* the powers of NGOs, international cooperation agencies and larger companies. Furthermore, network nodes and the entire network can change over time due to change in the power relationships among their members. In this process of change, new and different networks might be formed. Therefore, under such a new dominant power relationship based on democratic values, SMEs will be able to work hand-by-hand with local and international NGOs in a more influential and participatory way to develop sustainable markets at the local level and to access global sustainable markets.

Second, re-definition of power in networks is not free of struggle. Networks are a space of struggle since the members have their own interests and motivations. The expected shift from a top-down and paternalistic relationship of NGOs, international cooperation agencies and the government with SMEs will not happen without conflict and struggle. Particularly local NGOs, which are closer to SMEs, will be impacted by the rise of SMEs in the networks. Actors in the networks may abandon traditional perspectives and gain new capabilities to face successfully the re-definition of hierarchical power. Conflicts and disputes for power positions between NGOs and SMEs in the networks will be expected as long as SMEs gain in power position. Market dynamics can enhance struggles in the networks by motivating SMEs to use their resources to gain power.

Third, as both ecological modernization theory and network society theory posit, the power position of the national government will become less or even marginal in the networks of NGOs and SMEs. While the power of the national government decreases, the power of NGOs may become more central. Also, business actors, including SMEs, may gain a stronger power position. By consequence, the networks promoting sustainability of SMEs will be the result of NGOs and business relationship rather than of national government leadership. Obviously, national governmental agencies are also involved in promoting sustainability of SMEs. However, as the national government has showed to face structural limitations to handle environmental problems, low effectiveness of governmental agencies involved in promoting sustainable production for SMEs is expected.

2.5 Changing roles of NGOs in terms of changes in discourses and networks

Since this research sets out to describe changes in the relationships of NGOs and SMEs, it is useful to add to this theory chapter a brief consideration on identifying changes. Within the framework of this research, changes in the roles of NGOs will be measured in terms of changes in discourses and networks of NGOs.

Changes in discourse will be identified by analyzing the perspectives on environmental reform of SMEs that are advocated and promoted by NGOs. NGOs can potentially play influential roles in facilitating the exchange of sustainability perspectives from one place to another to foster environmental reform of SMEs. Such roles increasingly involve delivering and exchanging discourses containing particular perspectives about the liberal market, production and the environment, and shifts in the self-definition of NGOs from movement types to network types, and from charity deliverer to market player. NGOs not only hold an extraordinary vitality and dynamism in implementing discourses in practice, as Buttel (2001) argues, but they may also contribute with new perspectives to enrich traditional discourses and even create new discourses.

Changes in networks will be identified by analyzing shifts in the engagement of NGOs with larger companies, the joint collaboration of NGOs and SMEs, the emergence of new types of NGOs and the increase of power struggle in the networks as the result of power increase of SMEs. Such analyses will contribute to identify the changes of NGOs in terms of increasing diversification and flexibility of their roles towards collaboration with business (particularly SMEs) that are oriented to developing and accessing sustainable markets.

The theories, the analytical tools and the case domains are presented in a conceptual model that charts the research (Figure 2.1).

2.6 Relevance of the research

Despite the economic importance and recent growth in number of SMEs in Latin America, little attention has been paid to their sustainability. Sustainability of SMEs appears complex in the region since the environmental problems are interlinked with other relevant problems such as economic inequalities and political instability. Thus, it is imperative for the policy agenda of the region to explore strategies not only for improving the environmental sustainability of SMEs but

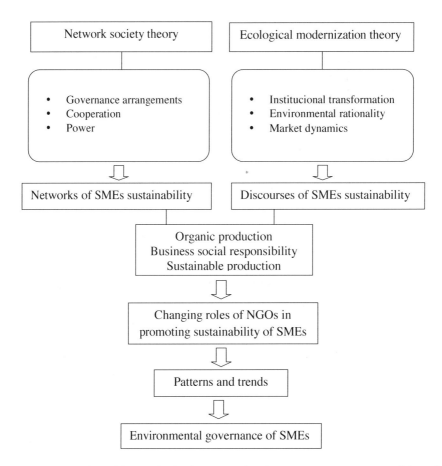

Figure 2.1. Conceptual model to study the changing role of NGOs promoting sustainability of SMEs in terms of changes in discourses and networks.

also to encourage the social and economic sustainability of SMEs. This thesis aims to contribute to policy making by researching such strategies.

Sustainability represents a challenge for SMEs in the region, particularly in Peru, and the design and implementation of sound policies of environmental reform of SMEs require the cooperation of several actors, with a significant role for NGOs. NGOs are non-state actors with a long history in the region. They have shown active and often leading participation in social movements confronting the business. However, during the last decades NGOs organized in international networks, including local and international NGOs, are working together with SMEs throughout the region to improve their environmental and social conditions. The remarkable feature of these developments is the fact that this is happening without the leadership of the national government and with a prominent position of NGOs *vis-à-vis* SMEs. The in-depth analysis of such international networks operating in Peru, and their (changing) discourses, will provide valuable inputs for the design of sound policies of environmental reform of SMEs in developing countries.

This thesis also contributes to the theoretical understanding of institutional change of NGOs in terms of their role, position and discourses in being potential agents of environmental reform of SMEs in developing countries. Some studies have been carried out on the role of environmental NGOs for environmental reform, but not specifically on local NGOs in the Latin American region, nor for local SMEs. Most studies focus on collaboration of international NGOs with larger companies. The innovative contribution of the research is the combination of network analysis and discourse analysis in identifying the main patterns and shifts in the relationships between NGOs and SMEs in dealing with sustainable production and consumption. Understanding such relationships is paramount to identifying the constraining and enabling factors in establishing sustainable practices in SMEs. According to Anheier and Katz (2004) an explicit and systematic focus on network analysis has not been common in global civil society research. Thus, the research makes substantial contribution to the empirical application of network analysis and discourse analysis to characterize the networks and discourses promoting sustainability of SMEs in a setting of a developing country. The research is also innovative in applying two of the major contemporary sociological theories – ecological modernization theory and network society theory – to frame such analyses.

Finally, in elaborating a set of case studies in the main domains promoting sustainability of SMEs in Peru, such as organic production, business social responsibility and sustainable production, the research provides fresh empirical evidence of environmental reform of SMEs. Thus, the thesis sheds light on the main pathways of environmental governance for the diversity of SMEs in Peru and the wider Latin American region.

Chapter 3.
Methodology

3.1 Introduction

This chapter presents in detail the concept of social network analysis (Section 3.2) and the concept of discourse analysis (Section 3.3), and their application to this thesis. Social network analysis and discourse analysis are the research analytical tools to be used in the thesis to operationalise ecological modernization theory and network society theory presented in Chapter 2 into an applicable set of key variables to guide the mapping of networks and discourses promoting sustainability of SMEs in Peru. The network mapping is based on the identification of key actors, cooperation strategies and power relationships in the networks under study. Similarly, the discourse mapping is based on the identification of basic entities, assumptions, agents and their motives, and key metaphors and their rhetorical devices of the discourses under study. The case selection (Section 3.5) and the data collection methods (Section 3.6) are also presented.

3.2 Social network analysis

3.2.1 Background

Network analysis is a useful tool to study social structure in which actors engage in more or less permanent and institutionalized interactions (Wellman, 1983, 1988). According to Mol (1995) network analysis has the advantage of combining both the structural properties of institutions and the interactions between actors constructing a network. Network analysis identifies the patterns of ties linking interdependent network members (Faust & Wasserman, 1994; Wellman, 1983, 1988). Granovetter (1983) highlights the importance of weak and strong ties in shaping network structure. According to this author, ties act as bridges between network segments such as groups in which most members are directly linked with each other; or, in the words of Wellman (1983), between clustered networks, facilitating macro organizational integration. The amount of ties in the network conditions the degrees of density of networks. Thus, networks can range from loosely bounded low-density networks to tightly bounded densely knit networks (Granovetter, 1983; Wellman, 1983). Ties provide the structural bases for cross-linkages among actors and, therefore, for cooperation (Wellman, 1983).

Networks involve power. Power can be expressed in the networks as function of actors based on their ability to control resources. Actors assume functions in the network, for instance, designing or operating assignments based on their ability to control knowledge, funds and other kind of resources valuable for the network members. Actors with strong influence and skill to control resources often are key actors, or 'nodes' in the networks. Indeed, a major characteristic of social networks is the power position of actors and the ways in which the patterns of ties allocates resources. In the words of Faust and Wasserman (1994), the ties among actors are channels to transfer resources, but the flow of resource distribution through the ties is not equal. According to Wellman (1983) the differential possession of scarce resources at the ties causes dependency relationships in the network. This author argues that unequal access to scarce resources may in turn increase the asymmetry of ties and it may lead to social structural change. Mol (1995) also highlights the importance of resources to structure the relations between different organizations in societal networks with respect to environmental reform. Therefore, networks are not only a space of cooperation among actors, but also of struggle for resources.

Network analysis seems to be an appropriate tool to study the complex influence that actors have on the environmental performance of SMEs and how actors and social structure limit, condition and enable the environmental reform of industries (Wattanapinyo, 2006). Similarly, network analysis is suitable to analyze the changing roles of NGOs, since NGOs structured in networks go across the nation-state, regions and continents to cooperate with SMEs. Most of the studies operationalizing ecological modernization theory (e.g. Chavalparit, 2006; Dieu, 2003; Frijns, Kirai, Malombe & Van Vliet, 1997; Hotta, 2004; Phoung, 2002; Wattanapinyo, 2006) have focused on the institutional context of business and their networks, but hardly any of them has primarily focused on NGOs and their networks. However, in the last years, social network analysis is increasingly being applied to study NGOs. A study that refers to networks and NGOs is that of Oosterveer (2005), who operationalized network society theory to study the dynamics of global food governance, and more specifically to identify new and innovative regulatory arrangements to deal with changing and new environmental and food safety concerns on a global scale. Besides official regulations and private certification schemes in governing food trade, Oosterveer highlighted the role of NGOs in introducing labeling arrangements, e.g. for organic farming and fair trade. Anheier and Katz (2003, 2004) showed the potential of network analysis for understanding global civil society by studying the connectedness of NGOs at global level, and Diani and McAdam (2003) used network analysis to study social movements.

3.2.2 *Application of network analysis in this thesis*

Network analysis is used in this thesis to systematically analyze the actual and potential roles of NGOs promoting sustainability of SMEs by studying the relationships of the actors involved in the networks. Especially, the analysis intends to capture the complexity of the relationships between NGOs and SMEs and between them and other actors in the network, identifying the main patterns and shifts. The analysis crosses the boundary of social and economic spheres since NGOs are usually considered part of social networks and SMEs are usually considered part of economic networks. By mapping the networks of NGOs promoting sustainability of SMEs it is possible to understand not only how network relationships constrain both the changes of roles of NGOs and the adoption of sustainable practices by SMEs, but also how network relationships enable those changes and adoptions.

Three key variables are identified to map the networks promoting sustainability of SMEs: (1) actors, (2) cooperation and (3) power. Actors involved in the networks of sustainability of SMEs are NGOs, SMEs, international cooperation agencies, larger companies and governmental agencies. Actors are organized in single networks or clustered networks with ties that go from local to global reach, beyond the national borders of Peru. Although analysis focuses on cooperation between NGOs and SMEs, cooperation of other actors with NGOs and SMEs is included. Cooperation of actors might be expressed in the networks by alliances, joint project implementations, coordinated actions, participation in common platforms and technical assistance oriented to environmental improvement of production of SMEs and to the position of SMEs in sustainable markets. Power relationships in the networks are identified in terms of the ability of certain actors to control resources. The analysis includes the identification of key actors, conflicts and other evidences of power struggle.

3.3 Discourse analysis

3.3.1 Background

Discourse analysis is a useful tool to study the shared ways of looking at the world called discourses. Discourse analysis is approached from two perspectives; one is mainly directed to the discursive process, the other to the discursive structure. While the first one focuses on communicative processes, the second one focuses on structure and coherence of discourses.

Environmental discourse, in the process perspective, is seen as a communicative process that places into debate belief systems, ideologies, norms and values in society, which are applied to human interrelations and to human relations with the environment (Mühlhäusler & Peace, 2006). It refers not only to the consensual perspective or viewpoint arrived at in such a communicative process, but also the competing and contentious discussions about, for instance, the environmental crises at the global level. In the process perspective, discourses are never static and rarely stable. At any one time there may be multiple and competing discourses in process of flux. Language is the means by which arguments about the relationship between humans and the nature are articulated and certainly the language one uses privileges certain perceptions and actions; expressing matters differently will privilege others. In the discursive processes, the ideas take on a specific meaning and inform particular practices, describing, explaining and justifying human-nature relationships (Benton & Short, 1999). In this perspective, environmental discourse is seen as overarching category with scarce coherence, including different speakers and addresses, and competing metaphors.

In the structural perspective, environmental discourses identify storylines and there can be several distinct discourses for a particular environmental problem (Hajer, 1995). Despite the fact that environmental discourses can be fragmented and contradictory with claims and concerns of a large variety of actors, a discourse, in this perspective, has specific structural characteristics with a certain level of consistence and stability. According to Dryzek (1997) a discourse rests on some common definitions, judgments, assumptions, and contentions. As a result, the way a discourse views the world is not always easily comprehended by those who subscribe to other discourses. The structure of a discourse is important as it conditions the way we define, interpret, and address environmental issues. In this thesis the analysis will mostly be directed to the discursive structure of discourses.

Environmental discourses have vastly increased in recent decades in response to worldwide awareness of environmental crises. There are quite a number of authors that use discourse analysis as a tool to frame concrete environmental problems and environmental politics. For instance, by studying the influence of ecological modernization on the regulation of acid rain in Great Britain and the Netherlands, Hajer (1995) analysed the policy discourse of ecological modernization and how this discourse affected policy making. Litfin (1994) analyzed the changes that have occurred in the international discourse on global ozone layer depletion in the 1980s. Environment is also gaining position in contemporary development discourses. For instance, authors such as Ebrahim (2005) use discourse analysis to study the natural resource management approach for rural development in India, putting special attention to the exchange of resources, the power relationships and organizational learning. Moreover, environmental discourses bring the concerns about environmental crises in wider political debates. A study that refers to this is done by Noble (2000), who assesed the degree of influence of NGOs and environmental discourse perspectives in the United Nations Conference on Environment and Development (UNCED)'s Declaration of Principles and Agenda 21.

3.3.2 Application of discourse analysis in this thesis

Discourse analysis is used in this thesis to systematically identify the main claims presented in the networks promoting sustainability of SMEs. In investigating these claims we focus on three critical issues for SMEs: market, environment and production. The analysis intends to identify the current discourses promoting sustainability of SMEs, find out their origins and foreseeing their future pathways. Moreover, our analysis intends to point out communalities and differences of discourses, to identify clues for cross fertilization, and to find out to what extent the identified discourses enrich or challenge traditional development discourses. Local perspectives may be as crucial as global ones in analyzing the discourses promoting sustainability of SMEs, so we will not only investigate the adoption of global perspectives at the local level but also the emergence of new perspectives enriching or challenging existing discourses from the local. By mapping the discourses promoting sustainability of SMEs it is possible to understand their constraining and enabling functions in establishing sustainable practices in SMEs.

As suggested by Dryzek (1997) four key variables are identified to map the discourses promoting sustainability of SMEs: (1) basic entities whose existence is acknowledged, (2) assumptions about natural relationships, (3) agents and their motives, and (4) key metaphors and their rhetorical devices. The four variables are structured in storylines for each identified discourse. Basic entities are views and elements that are acknowledged in the discourse. Entities can be, among others, political systems, ideologies and actors whose existence is acknowledged. Entities are the core claims of the discourse. Discourses imply assumptions in the relationships between different entities. The discourse can assume hierarchical relationships based on expertise, or virtue, or both; or equalitarian relationships. Hierarchy implies subordination to other entities. Other assumptions can be on cooperation or competition. Agents are individuals and collectivities that have agency capacities. Thus, agency is understood as having the capacity to act. Agency can be reserved for elites or granted to everybody. Motivations of agents can be material self-interest or public interest, among others. Key metaphors and their rhetorical devices highlight particular situations, issues or concerns in such a way that they express key perspectives of the discourse in a telling and understandable way.

3.4 Relationships of network analysis and discourse analysis

In analyzing the relationship between networks and discourses, this research distinguishes three main forms of relationships. First, network ties are important for the transfer of discourses. Network ties not only carry resources such as provision of funds and knowledge, but are also the means by which ideas and information flow through the network. Such non-material components are structured in discourse storylines, which flow between actors in the network. Second, the key actors and agents are linked with one another, and have a common reference. Key actors are elements of networks and agents are elements of discourses. While actors denote a function in the network, agents denote capacity to take action, and these two are clearly related. The position of NGOs as actors in the network, for instance, can be strengthened or weakened by the degree of agent capacity attributed to them is a discourse. Third, motivations and assumptions of agents affect the network relationships in terms of cooperation and power distribution. The establishing of ties in the networks is not enough for actors to cooperate. Actors have to be keen on working together, and this depends on whether they share discourses on how to improve environmental conditions of SMEs. Figure 3.1 presents the research analytical tools to study the relationships between networks and discourses promoting sustainability of SMEs.

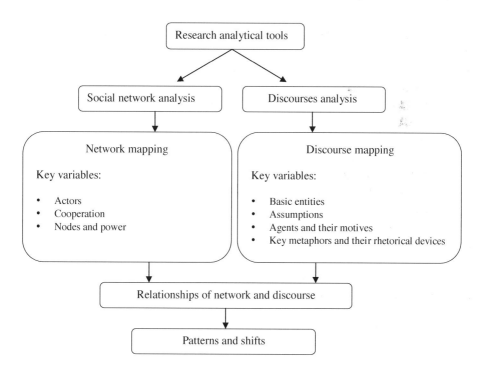

Figure 3.1. Research analytical tools to study networks and discourses promoting sustainability of SMEs.

3.5 Case selection

Case study research is adopted as the main strategy to carry out the gathering and processing of data into answers to the research questions. In order to guarantee sufficient variety on relevant data for the comparison and interpretation of cases, as Verschuren and Doorewaard (1999) recommend, three interrelated cases are selected: organic production, business social responsibility and cleaner production.

The case studies do not focus on NGOs as such but on networks and discourses in which, NGOs are major actors (in the networks) and major agents (in the discourses), and which include local SMEs (as an actor) and environmental reform of SMEs (as a basic entity). Most local SMEs tied to NGOs are producer associations, or agro-processing, garment, textile, leather and metal processing small enterprises, most of them supplying environmental friendly products to local and international markets. They can be urban-based SMEs (e.g. running small workshops and small industries) or rural-based SMEs (e.g. association of small producers, rural small industries, etc.). As NGOs and local SMEs, at least in part of the cases, are tied to international NGOs, international cooperation agencies and larger companies, those actors are also included in the cases. The overview of how the case studies are organized is presented in Table 3.1.

The three case study domains selected cover the most important networks and discourses of sustainability involving collaboration of NGOs and SMEs in Peru. Other domains of sustainability promotion involving NGOs and SMEs in the country are, for instance, green energy, ecotourism, fair trade and biotrade. Networks in those domains are not as well developed and large as the case study domains selected.

All three cases are characterized by significant changes from the 1970s until now, and therefore provide valuable clues to understand the changes of the discourses and networks promoting sustainability of SMEs. Furthermore, the large number of local and global actors involved in the

Table 3.1: Overview of case studies (see list of abbreviations for explanation)

Case study domains	Key NGOs	SMEs
Networks and discourses of organic production	RAE Peru Grupo Ecologica Peru ANPE	Small scale producer associations Small scale cooperatives
Networks and discourses of business social responsibility	CEDAL PERU 2021	Small scale enterprises
Networks and discourses of cleaner production	CER ITACAB Practical Action CONCYTEC CITEs IPES	Micro rural/urban enterprises Small rural/urban enterprises

three domains and the diversity of networks and perspectives are strong points with regard to the possibility of generalizing the results beyond the cases and beyond Peru.

3.6 Methods of data collection

Qualitative research methods are used to gather data for the case studies. Data for qualitative research can be distinguished, according to how they are acquired, into primary and secondary data.

The case studies are elaborated based on primary and secondary data collected. Primary data collection included the application of in-depth interviews with semi-structured questionnaires, formal and informal meetings, field observations, and seminar/workshops participations. Interview questionnaires were adapted to the specific respondents, but for each of the domains a specific general guideline was used to structure the interviews. Those guidelines are differentiated for local NGOs (Appendices 3a, 3b and 3c), for international NGOs (Appendix 3d) and for SMEs (Appendix 3e). The questionnaires were designed to collect data and information for the aforementioned key variables of networks and discourses. In each NGO selected, a number of key informants were interviewed face-to-face. International NGOs interviewed were selected on the basis of their links with the key local NGOs identified. Extra interviews were done at local levels to SMEs representatives and relevant government officials and experts at the national level. A complete list of interviewees is presented in Appendix 1. The field data gathering in Peru was conducted in the departments of Lima, Huancayo, Arequipa and Trujillo from April 2006 to January 2010, including the exploratory and main fieldwork phases. Secondary sources of data gathering included reports, publications and websites of NGOs, SMEs and governmental agencies, and scientific journal articles and books.

In total 28 persons of local NGOs, international NGOs, SMEs and the national government including programme coordinators, managers and directors were interviewed. They usually were interviewed, depending of the relevance of the information, more than one time, and in the case of the key NGOs they were tracked during the entire fieldwork period.

Chapter 4.
Three global mechanisms for greening SMEs

4.1 Introduction

Business social responsibility, organic production and sustainable production are three mechanisms of global scope that aim to include social and environmental concerns in production and markets. As explained in the Chapter 1, the domains selected in this research to investigate the roles of NGOs in promoting sustainability of SMEs correspond with these three mechanisms. This chapter provides an overview of the three domains. For each domain the concept, the development and trends, and the key agents involved are described. To provide a global context for the chapters to follow, the overview gives particular attention to the international level.

4.2 Organic production

4.2.1 Concept

Organic production, also referred to under the name of organic farming and organic agriculture, is a widely respected process in agricultural production that aims to improve soil fertility, nutrient cycling, and food security, and reduce the use of veterinary medicines. A central element of organic production is the efficient use of on-farm and local resources like farmyard manure, indirect crop protection and local seed (Vogl, Kilcher & Schmidt, 2005). Organic production intends to overcome the negative environmental impacts such as global warming, reduction in biodiversity, soil degradation and ground water pollution caused by agrochemical-based agriculture (Halberg, Alrøe, Knudsen & Kristensen, 2006). Especially in developing countries, organic production not only contributes to improving the environment but also aims at strengthening socio-economic development.

Organic production is embedded in local cultures, including their knowledge, ethical values and beliefs (Yussefi & Willer 2003). According to Benzing (2001) traditional agriculture provides useful models of sustainable agriculture and its achievements, especially in Australia, India, Central and South America, have served as foundations to develop modern organic production. For instance, the traditional agriculture of Central and South America has been central in the development of the concept of agro-ecology (Altieri, 1999; Altieri & Nicholls, 2000). Peru being one of the world's five centres where agriculture has originated, traditional agriculture presents particular features such as the vertical control of 'ecological floors' (one of the ways Andean societies organize the supply of goods, including food, because of the steepness of the Andes), diversification of crops and indigenous farming technologies (Altieri, 1995; Espinoza, 1987; Murra, 1975; Rostworowski, 1999). Furthermore, organic production in Latin America, particularly in Peru, not only encompasses traditional farming practices but also a dynamic movement of farmers and organizations (Altieri & Nicholls, 2000; Wezel, Bellon, Doré´, Francis, Vallod & David, 2009). These particularities show that organic agriculture has local and regional distinctiveness, although it is a global phenomenon (Holt & Reed, 2006; Kristiansen, Taji & Reganold, 2006).

4.2.2 Developments and trends

After the 1960s the organic production movement underwent significant change. The appearance of Rachel Carson's 'Silent Spring' in 1972, the Club of Rome's 'Limits to Growth' in 1972 and Schumacher's 'Small is Beautiful' in 1974 marked a turning point. One of the results was that the organic and environmental movements came together. A major landmark was the formation of the International Federation of Organic Agriculture Movements (IFOAM) in 1972 (Kristiansen, Taji & Reganold, 2006). During the last 25 years, organic agriculture expanded from a movement with political, philosophical and religious motivations to a movement with economic motivations. This is shown in the responses against the introduction to Genetic Modified (GM) crops at country level because such technologies are seen by organic agriculture embracers as a threat for traditional agriculture and crops (Altieri & Nicholls, 2000; Halberg *et al.*, 2006; Kristiansen, Taji, & Reganold, 2006). These characteristics provide sociological and political relevance to organic production (Holt & Reed, 2006).

In the last two decades organic production has achieved formal political and legislative recognition at national and international level. Together with governments and other stakeholders, the organic production sector has developed standards and legislation on organic farming, which include rules for processing, trading, monitoring, and certifying agricultural produce. Examples of such standards are the European Council Regulation on Organic Farming No. 20292/91, the IFOAM Basic Standards, and the US Organic Food Production Act.

Organic production is currently distributed in about 100 countries and the area is increasing. When it comes to certified organic land Australia, Argentina and Italia are located at first place. The regions with largest areas of organically managed agricultural land are Oceania (12.1 million hectares), Europe (8.2 million hectares) and Latin America (8.1 million hectares) (Willer & Kilcher, 2010). About one-third of the world's organically managed agricultural land (12 million hectares) is located in developing countries. Most of this land is localized in Latin America. In developing countries organically managed agricultural land is owned mostly by small scale organic producers. According to Willer and Kilcher (2010), the countries with the highest number of organic producers are India (340,000 producers), Uganda (180,000) and Mexico (130,000).

Organic production and consumption has increased over the last decade. Although many organic products are consumed locally, global trade of organic products, especially between North and South, is a growing reality (Halberg *et al.*, 2006; Vogl, Kilcher & Schmidt, 2005). In the 1980s organic production experienced explosive growth. In the major markets for organic produce – North America, western Europe and Japan – organic produce began to appear in supermarkets in the early 1980s (Holt & Reed, 2006). In the late 1990s organic production strongly became more rationalized and market-driven. In the 1990s, and into the new millennium demand and production continued to grow exponentially around the world, often at 20-30% per year. Most consumption takes place in western countries. The global organic market involves about US$ 23 billon annually (Holt & Reed, 2006). Europe and North America represent the major markets for certified organic products, accounting for roughly 97% of global revenues (Kristiansen, Taji & Reganold, 2006). Products traded are coffee, tea, cocoa, spices, cane sugar, tropical fruits, and beverages, as well as seasonal fresh produce (Halberg *et al.*, 2006). However, the average organic production per country is less than 6% of the total agricultural production, so compared with the conventional production the share of organic production is modest (Holt & Reed, 2006). Another development of the two decades has been the resurrection of the farmers' market where producers sell goods directly to consumers (Kristiansen, Taji & Reganold, 2006).

Global markets impact organic production by pushing farming to rationalization. Rationalization, in this context, refers to specialization and enlargement of farms, increasing capital intensification and marketing, and a strong orientation on export. These developments represent threats and opportunities for organic production. In most national organic sectors across the planet, agribusiness has now moved into production of large scale. This has created barriers for small-scale producers to access global markets (Holt & Reed, 2006). International trade of organic production is mostly controlled by medium and large enterprises, not by small scale enterprises (Halberg *et al.*, 2006). Nonetheless, expanding markets go together with broader recognition of organic farming principles and may provide new options for organic producers. However, there is a need to develop policies for organic production that take into account local food security, environmental impacts, and socio-economic impacts. Locally traded organic production provides social and economic value such as social cohesion and expanding employment in rural areas (Halberg *et al.*, 2006). In sum: organic production has gained a place in global markets, but it is still marginal and has low power against conventional agribusiness trade, and social and environmental challenges of organic production remain (Kristiansen, Taji & Reganold, 2006).

In recent years, international cooperation agencies and international NGOs supporting SMEs have adopted and promoted new approaches to rural development, which also affect organic production. NGOs are shifting from charity to partnering, from projects to programmes, and from service delivery to advocacy. For instance, in a study about the roles of NGOs in Peru and Bolivia, Bebbington, Rojas and Hinojosa (2002) conclude that local NGOs are not succeeding in poverty reduction and rural development, and suggest that international cooperation should shift to a more market-based approach.

4.2.3 Agents

Producer associations, NGOs, companies and national governments perform roles promoting organic production and trade at the international and national level (Bridge, O'Neill & Martin, 2009). The key global agent is the International Federation of Organic Agriculture Movements (IFOAM). IFOAM is the worldwide umbrella organization for the organic movement, uniting more than 750 member organizations in 116 countries, including small scale producer associations. IFOAM actively participates in international agricultural and environmental negotiations with the United Nations and multilateral institutions to further the interests of the organic agricultural movement worldwide (IFOAM website, 2011). Other agents are NGOs. Increasingly NGOs, particularly international NGOs, are taking an active part in promoting and supporting SME development in the organic sector. For instance, Dutch development aid organization CORDAID supports SME development by improving the enabling environment that conditions the development of sustainable agricultural value chains and the sustainability of well-organised agricultural chains, building on experiences of organic and fair trade chains (CORDAID website, 2011). Two other Dutch development aid NGOs, HIVOS and ICCO, help small scale producers in developing countries to access local and international markets with an organic and fair trade quality mark. They collaborate with larger companies and local NGOs to develop organic and fair trade certification and standards (HIVOS Annual Report, 2011; ICCO Annual Report, 2011). NGOs such as OXFAM America also work closer to grass root organizations and academia around the world to promote self-sufficiency and food sovereignty (Cohn, Cook, Fernández, Reider & Steward, 2006). Additionally, international cooperation agencies, including the Food and Agriculture Organization of the United Nations (FAO), the international fund for agriculture (IFAD), the Danish Research Center for Organic Farming (DARCOF), the Belgian Technical Cooperation (ADG), the Danish Agency for Development Assistance (DANIDA) and the Research Institute of Organic Agriculture (FiBL) are also active in promoting the globalization of organic production (Halberg *et al.*, 2006). Also, national governments in western countries are promoting organic production. National governments in developing countries are increasingly showing interest in implementing regulations to support exporting certified organic production (Kristiansen, Taji & Reganold, 2006).

At the Latin American regional level, the main agents are the Latin American agroecological movement (MAELA) and the IFOAM Regional Group for Latin America and the Caribbean (GALCI). In Peru, key agents are the Ecological Agriculture Network of Peru (RAE Peru), Grupo Ecologica Peru and the National Ecological Producers Association (ANPE). Local agents will be described in more detail in Chapter 5.

4.3 Business social responsibility

4.3.1 Concept

Business social responsibility is the process whereby a company on a voluntary basis integrates social and environmental concerns in its business operations and in its relations with their stakeholders. The concept is equivalent to 'corporate social responsibility'. In international literature the latter term prevails (Prieto-Carrón, Lund-Thomsen, Chan, & Muro, 2006), but in this thesis we will use the term business social responsibility because of the Latin-American context of research. In Latin-America the term business social responsibility is more common, especially if it is referred to local companies (Peinado-Vara & De la Garza, 2007), and corporate social responsibility usually refers to large multinationals and big corporations (Casanova & Dumas, 2010).

Business social responsibility comes with globalization and the increasing concern about social and environmental issues in production and trade. Its principles revolve around the following issues: (a) companies have a responsibility for their impact on society and the natural environment, sometimes beyond legal compliance and the liability of individuals; (b) companies have a responsibility for the behaviour of others with whom they do business, for instance within supply chains; and (c) businesses need to manage their relationship with wider society, whether for reasons of commercial viability or to add value to society (Blowfield & Frynas, 2005).

To improve business operations, business social responsibility invokes, among other means, new standards (Schwalb & Malca, 2004), for example, standards for reporting, measuring and auditing. Well-recognized standards are SA 8000 (social accountability), Global Reporting Initiative and ISO 26000. Other norms applied also are the ISO 9000 series (for quality management) and the ISO 14000 series (for environmental management). Also, eco-efficiency and industrial symbiosis (cooperation among industries in improving environmental performance) are recently seen as ways to ground business social responsibility in companies. Indices and indicators for business social responsibility have been developed by the Instituto Ethos of Brasil, the NGO Business for Social Responsibility and the World Resource Institute, among other organizations (Schwalb & Garcia, 2004). Proponents of business social responsibility claim that it not only improve the social and environmental performance of a company, but also helps to increase its market share and improve its access to global markets (Blowfield & Frynas, 2005; Schwalb & Malca, 2004).

4.3.2 Developments and trends

Major landmarks in the development of the concept of business social responsibility are the Ethical Codes for Multinational Companies of the Organisation for Economic Co-operation and Development (OECD) drafted in 1976, the Tripartite Declaration of Principles concerning Multinational Enterprises and Social Policy (MNE Declaration) of the International Labour Organization (ILO) in 1977, the launching of the report 'Our Common Future' by the Brundland commission in 1987, the United Nations Conference on Environment and Development (UNCED) in 1992, the elaboration of most of the business social responsibility norms and indicators from 1996 to 2000, the United Nations Global Compact launched in 2000, and the EU Green Paper elaborated in 2001 (Hall, 2010; Schwalb & Malca, 2004; Schwalb & Garcia, 2004). Recently, in 2010 the ISO 26000 is launched to provide organizations with guidance concerning social responsibility and an International Integrated Reporting Council (IIRC) is established (BSR website, 2012).

In developing countries, business social responsibility initiatives have particularities. Compared to developed countries, more emphasis is put on the improvement of working conditions, human rights and contribution to poverty alleviation of communities (Prieto-Carrón *et al.*, 2006). According to Agüero (2002), the rise of business social responsibility in Latin America has to do with three interrelated factors: (1) social pressure from the bottom; (2) change of vision among business leaders; and (3) developments in the theory and practices of business management. Additionally, the persistent signs of poverty and inequalities in the 1980s and in the 1990s in the context of less competent states and more powerful companies might have also played a role. Those factors contributed to shape business social responsibility of companies in Latin America with a philanthropic orientation (Agüero, 2002).

Regarding Peru, business social responsibility has evolved as a more active participation of business people in the social concerns of the country, especially after the period of political violence in the 1980s. For instance, Peru 2021, one of the main NGOs promoting business social responsibility in Peru, was founded after that period with the aim to develop a national vision for the country, pointing out a more committed role of business in development issues (Agüero, 2002). During the last decade local NGOs have also started to be engaged with companies implementing initiatives of social responsibility (Caravedo, 2003). However, the adoption of business social responsibility by larger companies in Peru, especially by the extractive mining and oil multinationals, have usually been accompanied by social conflicts that questioned their social legitimacy (Schwalb & Malca, 2004). According to Caravedo (2003) embracing the concept of business social responsibility in Peru is still a challenge for most of the national companies, most of them being SMEs.

4.3.3 Agents

Agents involved in the promotion of business social responsibility are companies, international cooperation agencies and NGOs. Key global agents are the World Business Council for Sustainable Development, WBCSD (founded in 1990), the European business network for corporate social responsibility (founded in 1995), the World Resource Institute (WRI), the World Bank, the United Nations Global Compact, the Foundation for Research on Multinational Corporations (SOMO), OXFAM and the Interchurch Organisation for Development Cooperation (ICCO) (Schwalb & Malca, 2004; UN Global Compact website, 2011).

Many NGOs involved in this domain are organized at the international level and have assumed a 'watchdog' role and collaborative engagement with business actors to further social responsibility in society. OXFAM Novib is affiliated to OXFAM International, a confederation of 14 like-minded organizations active in almost one hundred countries around the world. The OXFAM confederation is a development and campaigning organization advocating at the global level the rights of people living in poverty. One of its core activities is to lobby governments, institutions and businesses to make socially responsible choices and take action for climate change adaptation in developing countries. In Latin America, OXFAM Novib has set up and financially supports the Red Puentes, a regional platform of NGOs to promote business social responsibility. Currently, OXFAM International is subject to change. Although OXFAM International keeps operating in Latin America, OXFAM Novib decided to withdraw from this region in 2011. Actually, OXFAM Novib has reduced the number of countries where it works from 70 to 34. However, by partnering with other global organizations OXFAM Novib keep encouraging the private sector to take on board corporate social responsibility in the region. Another development in the last two years has been the setting up of OXFAM India and OXFAM Mexico, and in the upcoming years it is expected the seeting up of national OXFAM affiliates in Brazil and South Africa (OXFAM-NOVIB Annual Report 2011).

SOMO conducts research and other activities focusing on the policies and conduct of multinational corporations in an international context. By collaborating closely with other NGOs and their networks, SOMO discloses pressing issues in the supply chains of brand and retailing companies. For instance, SOMO is part of the 'Clean Clothes Campaign' in the garment & textile sector and helps Red Puentes in campaigning and lobbying in Latin America and Europe. On behalf of Red Puentes, SOMO has been participating in the ISO Working Group that developed the ISO 26000 standard (see below) (SOMO Annual Report 2009).

ICCO focuses on the development of local and regional markets. ICCO supports local producers and market chains with start-up grants, loans and guarantees. ICCO cooperates with producer organizations, co-operatives, companies, governments and other NGOs. ICCO has established the Fair Climate Fund in 2009 to buy and sell the carbon credits generated in their projects supporting agri-SMEs on the voluntary Dutch and European market. The year 2010 has been a turning point for ICCO in its way of working, as it turned to a more decentralized way of working by establishing regional offices around the world. Since 2010 ICCO has also been keen to engage with socially responsible companies seeking to make development cooperation more effective (ICCO Alliance Annual Report 2010; ICCO website, 2011).

At the Latin American level, the main agents are the Inter-American Development Bank (IDB), AVINA, the Instituto Ethos, the Red Puentes International and the Forum Empresa. It is worthy to mention that the Business for Social Responsibility, the main organization in the US promoting business social responsibility, played a key role in providing advice and being a model for Latin American NGOs promoting business social responsibility (Agüero, 2002). In Peru, key agents are Peru 2021, the Red Puentes Peru and the Pacific University. Local agents will be described in more detail in Chapter 6.

A recent effort involving key global, regional and local actors from 99 ISO member countries has been the development of the standard ISO 26000: 2010. The five-year development process involved among others industry, government, labour, consumers, NGOs and research representatives from developed and developing countries. ISO 26000: 2010 is an ISO International Standard giving guidance on social responsibility to all types of organizations, regardless of their size or location (ISO website, 2011). However, the new standard has been criticized by SME organizations for not meeting their needs (Area Minera website, 2012; NORMAPME website, 2012), especially considering that the application of standards and guidelines of social responsibility is challenging for SMEs (Hall, 2010).

4.4 Sustainable production

4.4.1 Concept

Sustainable production means the efficient management of resources at all stages of value chains of goods and services. Sustainable production encourages the development of processes that use fewer resources and generate less waste, including hazardous substances, while yielding environmental benefits and frequent productivity and economic gains. Sustainable production also encourages capturing and reusing or recycling valuable resources, thereby turning waste streams into value streams (Beaton & Perera, 2012; Ekins & Lemaire, 2012).

Two of the central ideas of sustainable production are decoupling and leapfrogging. Decoupling refers to the process of reducing the resource intensity of, and environmental damage relating to, economic activities; in other words, the decoupling of economic growth from environmental degradation. Leapfrogging expresses the fact that societies do not need precisely to imitate each other in their processes of development. They can adopt modern technologies without following the development trajectories undertaken when those technologies were not available (Beaton & Perera, 2012).

Cleaner production and appropriate technology are two key approaches under the umbrella concept of sustainable production. Cleaner production is a worldwide approach for environmental pollution control and prevention in industry. On the one hand, inefficient production methods lead to wasting of resources, economic losses, poor working conditions and environmental pollution. On the other hand, production processes that are more energy efficient, use fewer resources, and re-use waste materials not only reduce environmental impacts but may also reduce economic costs. Pollution prevention therefore can be a more attractive approach for industries than treatment of waste by end-of-pipe measures that add on costs. Cleaner industrial production must be concerned not only with reducing levels of toxic and hazardous waste emissions but also with life cycle analysis, including the whole production and consumption process to achieve more efficient use of resources (OECD, 1995).

Cleaner production measures have proved to be particularly attractive for SMEs in developing countries (OECD, 1995). Several studies, mainly in Asia, have been already done on cleaner production as industrial restructuring strategy of SMEs (Evans & Hamner, 2003; Sonnenfeld, 2000). However, SMEs are constrained in implementing cleaner production by various attitudinal, organizational, technical and economic barriers (Frijns & Van Vliet, 1999). Although cleaner production offers opportunities for 'win-win solutions', where environmental improvements go hand-in-hand with economic benefits, its implementation often remains difficult and will not happen overnight. Cleaner production is not simply a matter of applying new technologies but requires readjustment and rethinking throughout the company (Christie, Rolfe & Legard, 1995). So, cooperation, involvement and commitment of management and employees are essential to implement cleaner production methods (Zwetsloot & Geyer, 1996). Despite the potential saving in costs, cleaner production is restrained by financial obstacles, for instance, low prices for raw materials, low waste disposal charges, and lack of funding. Successful introduction of cleaner production methods for environmental improvements require better coordination of policy efforts and organization of small scale enterprises (Frijns & Van Vliet, 1999).

The approach of appropriate technology refers to environmentally sound technology that puts an emphasis on small scale solutions, use of intensive labor, efficient use of locally available energy and local control (Schumacher, 1983). Therefore, sustainable production not only focuses on eco-efficiency and the transfer of modern large scale technologies but also on saving energy and resources based on the adjustment of technology to the local context (NCAT website, 2011; Practical Action website, 2011). Appropriate technology is equivalent to intermediate technology. Appropriate technology might be the result of the improvement of a traditional technology, the adaptation of an advanced technology to local conditions or the development of a new 'tailor made' technology. Appropriate technology is intended to tackle the need of technology of poor people in rural and urban areas based on a regional focus and on decentralization. The typical sectors of application of appropriate technology are building materials, garment, farming tools, post harvesting methods and basic industrial transformation of agricultural products (Schumacher, 1983).

4.4.2 Developments and trends

The evolution of sustainable production is related to a number of developments. The 1992 United Nations Conference on Environment and Development in Rio de Janeiro (the Earth Summit) is the early landmark for many efforts to promote sustainable production at the regional and national levels. Agenda 21, the action plan for sustainable development adopted at the summit, called for action to promote patterns of consumption and production that reduce environmental stress and will meet the basic needs of humanity (Beaton & Perera, 2012). In 1993, the government of Norway at the United Nations Conference on Trade and Development (UNCTAD) organized a workshop on the transfer and development of environmental sound technologies. In the same year, the government of Colombia and the United States sponsored preparatory meetings on technology transfer, cooperation and local capacities. In April 1994 the Agenda 21 process called for the transfer of environmentally sound technology, cooperation and capacity building (OECD, 1995). In 1995 the OECD held a workshop for development assistance and technology cooperation for cleaner industrial production in developing countries (OECD, 1995). The main landmark in the promotion of cleaner production is the launching of the National Cleaner Production Centres (NCPCs) programme in 1995 by the United Nations Industrial Development Organization (UNIDO) and the United Nations Environment Programme (UNEP) (UNEP website, 2011; UNIDO website, 2011). In 2002 at the Johannesburg World Summit on Sustainable Development (WSSD) a 10-Year Framework Programme on Sustainable Consumption and Production (10YFP) was approved to promote the shift to sustainable production and consumption. 10YFP was followed by the Marrakech Process, which developed various mechanisms, including regional consultations, Task Forces, and dialogues with different stakeholders, in order both to refine the concept of sustainable production and to show how it could be made operational in very different countries, economic sectors and cultural contexts (Beaton & Perera, 2012).

Compared to cleaner production, the attention to appropriate technology has declined over the last two decades. For instance, two of the main organizations promoting it, the German Appropriate Technology Exchange (GATE) and the Holland's Technology Transfer for Development (TOOL), are no longer in operation. However, Intermediate Technology Development Group (ITDG), the organization founded by Fritz Schumacher in 1969, is still in operation and in the last years appropriate technology seems to re-emerge again. This re-emergence has to do with the rise of the open source appropriate technology (OSAT) such as Apropedia. This new wave of appropriate technology has benefited from the opportunities now offered by the Internet. OSAT refers to technologies that provide for sustainable development while being designed in the same fashion as free and open source software. According to Pearce (2012), OSAT is made up of technologies that are easily and economically utilized from readily available resources by local communities to meet their needs and must meet the boundary conditions set by environmental, cultural, economic, and educational resource constraints of the local community.

4.4.3 Agents

Agents promoting sustainable production intervene at global, regional and national level. Key global agents of cleaner production are the United Nations Environment programme (UNEP), the United Nations Industrial Development Organization (UNIDO), the World Bank, the Inter-American Development Bank (IDB), the Global Environment Facility (GEF) and the Swiss State Secretariat for Economic Affairs (SECO). They play key roles in establishing policies in developing countries to support cleaner production diffusion, making pollution prevention an integral part of their technical assistance activities, enhancing the capacity development to manage technological change and facilitating access to information and the technological innovation, diffusion and implementation (OECD, 1995). UNEP is an international organization that coordinates United Nations environmental activities and assists developing countries in implementing environmentally sound policies and practices. UNIDO is the specialized agency of the United Nations that promotes industrial development for poverty reduction, inclusive globalization and environmental sustainability. UNEP and UNIDO have worked on promoting cleaner production in SMEs for several years but since 2011 they are working together to further cleaner production under the umbrella of sustainable consumption and production (UNEP website, 2011; UNIDO website, 2011)

The World Bank is an international financing institution consisting of five organizations. One of them is the International Finance Corporation (IFC). IFC provides loans and technical assistance to stimulate private sector investment in developing countries. IFC is interested in helping clients to take full advantage of cleaner production techniques by offering advice to optimize facility and process design, identify retrofit opportunities and explore alternative energy and other resource efficiencies. SMEs across a range of sectors and regions are part of their clients (IFC website, 2012; World Bank website, 2011).

IDB is the largest source of development financing for Latin America and the Caribbean. In 1993 IDB established the Multilateral Investment Fund (MIF) to finance small projects that promote innovative development of the private sector and strengthen SMEs. The projects include energy efficiency technologies and practices, and other climate change adaptation strategies (IDB website, 2011; MIF website, 2011).

SECO's aim regarding international cooperation is to integrate partner countries into the world economy and promote their sustainable economic growth, making an efficient and effective contribution to reducing poverty. In close collaboration with the Swiss Agency for Development and Cooperation (SDC), SECO has been financing the implementation of Cleaner Production Centers (CPCs) in Africa, Asia, and Latin America (SECO website, 2011).

The Global Environment Facility (GEF) is a financial organization integrating 182 member governments established in 1991. GEF provides grants to developing countries and countries with economies in transition for projects related to biodiversity, climate change, international waters, land degradation, the ozone layer, and persistent organic pollutants. GEF is today the largest funder of projects to improve the global environment. GEF works in close collaboration with 10 international agencies, including the UN Development Programme, the UN Environment Programme, the World Bank, the UN Food and Agriculture Organization and the UN Industrial Development Organization. The allocation of grants to implement cleaner production centres in order to improve environmental conditions in SMEs is one of its priorities (GEF website, 2011).

Regarding appropriate technology, the principal worldwide agents are ITDG (now Practical Action) and the International Development Research Center (IDRC). Other key agencies are the National Center for Appropriate Technology (NCAT) in USA and, the Centre for Appropriate Technology (ICAT) in Australia. During the last years new pioneer organizations promoting appropriate technology at global level have emerged, including Appropriate Technology Collaborative (ATC), Catalytic Communities (CatCom), Appropriate Infrastructure Development Group (AIDG), Digital Green, Engineers Without Borders, Kopernik and Village Earth (Pearce, 2012; Zelenika & Pearce, 2011).

Agents operating at a regional level on cleaner production development are the Organization of American States (OAS) and the Spanish Agency of International Cooperation (AECI). The OAS brings together all 35 independent states of the Americas and constitutes the main political, juridical, and social governmental forum in that hemisphere. One of the main tasks of OAS is to formulate and execute technical cooperation projects, including follow-up of the mandates of the Summit of the Americas on Sustainable Development. Projects include the promotion of sustainable technologies to improve the conditions of SMEs in the region (OAS website, 2011).

AECI works in 50 countries, including Latin America countries. One of its main aims is to promote productive activities by enhancing business capacities, developing markets, connecting SMEs to markets and enhancing innovation and technology. During the last years the support of a network of Centers of Technological Innovation (CITEs) in Latin America has been one of the projects funded by AECI (AECI website, 2011).

At the local level, key agents promoting cleaner production during the last years in Peru are the NGOs CER and IPES, the (inter)governmental agencies ITACAB and CONCYTEC, and the public-private organization CITEs. The main organizations promoting appropriate technology are Practical Action Peru (a local NGO linked to Practical Action UK), the Catholic University's GRUPO and the National University of Engineering. Local agents are described in more detail in Chapter 7.

Chapter 5.
Organic production

5.1 Introduction

The 'green revolution' has mainstreamed large scale agro-industries, farming of single crops, and the use of synthetic fertilizers and pest controls in worldwide food production. Although this development has made possible the increase in productivity, it has brought concerns for its negative environmental impacts and for eroding worldwide traditional farming. Contesting the green revolution, organic production is a different way to approach food production that intends to produce higher food quality, with less external inputs and less external side-effects. Due to increasing market demand, organic production is expanding, both in terms of small scale organic farms and larger organically managed farms. As a result, organic production is showing points of convergence with modern agriculture such as the production on larger scales and in monocultures.

In developing countries the rise of the organic market has become an opportunity for small scale producers to improve their economic conditions, while natural resources such as crop diversity and water are protected and climate change is mitigated. Therefore, stimulating the expansion of organic food enterprises and local organic markets in developing countries contributes to strengthening sustainable rural development. In Peru organic production is in many respects in line with ancestral farming practices. As a result, local small scale producers have converted faster to organic production in comparison to local large scale producers. Peru is the second largest country in Latin America and seventh in the world in number of small scale organic producers, with about 36,000 small scale organic producers (Willer & Kilcher, 2009). Peru is also the country with most agricultural land in Latin America and sixth in the world in organic farming area (Willer & Kilcher, 2010). Most local organic production in Peru is produced for foreign markets: approximately 98% of total certified organic production is exported to foreign markets representing 8% of national agricultural exports and only 2% is commercialized at domestic market, according to Centro IDEAS (Boletin de IDEAS, 2011). From the amount commercialized at domestic market, 44% is commercialized at local markets, 43% directly sold by producers to consumers and 11% at Bioferias (Alvarado, 2002; RAAA, 2007). The selling of organic vegetables at domestic supermarkets represents less than 0,6% of the total vegetable selling, although it growths 7% each year (Cavero, 2007).

Small scale producers are the main actors in agricultural production in Peru. Overall small scale agriculture, predominantly 'conventional' production, represents 90% of all agricultural production in Peru. From this amount, 3% correspond to certified organic producers (Boletin de IDEAS, 2011). According to FAO (2009) small scale producers supply about 80% of their produce to local markets in Latin American countries. Small scale producers are categorized as having less than 20 hectares, while medium-sized producers have from 20 to 49.9 hectares (Torres, 2004).

In this research organic production, organic agriculture, agro-ecological and ecological production are used with an equivalent meaning (see Chapter 4). Also, small holder, small farmer and small scale producer refer to the same subject. Additionally, small scale producers, including producer associations, cooperatives or single small agri-businesses are included under the 'umbrella' name of small scale enterprises (see Chapter 4). According to the Peruvian law of organic production, producers are named Agrarian Production Units.

Organic production involves in Peru a variety of NGOs. They have played central roles in supporting the rise of small scale producers as legitimate economic actors and the development of organic market in Latin American food, particularly in Peru. NGOs with a long history – which are called 'traditional' NGOs for the purpose of this research –, are part of the networks and discourses promoting organic production. In this research conventional NGOs refer to NGOs founded in the 1980s that currently perform as market facilitators bringing to small scale producers basic competences in organic production. Usually this type of NGOs consists of committed professional and experts formally educated at universities. Examples of conventional NGOs are among others Centro IDEAS, CEAR, IDMA and AEDES. NGOs performing as market intermediaries of small scale organic producers supplying to markets are called 'market' NGOs in this research. NGOs intermediating in local markets are for instance Grupo Ecologica Peru and Taller. NGOs intermediating local organic production towards international markets are for example; PIDECAFE and CANDELA Peru. Finally, NGOs rooted at communities, usually led by producers – which I call 'producer' NGOs – are also part of the organic production approach. This last type of NGOs has appeared as answer to questions of representativeness and coordination of small scale organic producers, widely spread throughout in Peru. An example of a producer NGO is ANPE.

The chapter is organized in six sections. Section 5.2 presents the network analysis of organic production and describes in detail the main networks identified in Peru: the agro-ecological network, the organic market network and the ecological farming network. The section includes an overview of the Bioferias, the organic farmers' market in Peru. Afterwards, the main patterns, challenges and trends of the networks, and the main conclusions of the network analysis are presented. Section 5.3 presents the discourse analysis of organic production and describes in detail the main discourses identified in Peru: the discourse of market adaptation, the discourse of market access and the discourse of market democratization. Subsequently, the main conclusions of the discourse analysis are presented. In Section 5.4 the three discourses on organic production are allocated in a two dimensional policy realm. Section 5.5 presents the major patterns and trends of the networks and the discourses of organic production. Finally, the main conclusions of the chapter are presented in Section 5.6.

5.2 Network analysis of organic production

Before analysing in depth the key network members, the network relations and the main network changes and challenges, the three networks identified are introduced briefly. These networks are named: (1) 'the agro-ecological network'; (2) 'the organic market network'; and (3) 'the ecological farming network'. They are the main networks in Peru stimulating small scale producers to supply organic food to local and international markets. Each network includes a particular set of local, regional and global actors. Agro-ecological NGOs are key actors in each of the three networks. They work in close cooperation with organic producers.

The first network to be analyzed is the agro-ecological network. The key actor in this network is the Ecological Agriculture Network of Peru (RAE Peru). The second network to be analyzed is the organic market network. The key actor in this network is Grupo Ecologica Peru. The third network is the ecological farming network. The key actor in this network is the National Ecological Producers Association (ANPE). The three networks have different approaches to organic production. The agro-ecological network is more directed to supporting producers to learn basic farming and business skills of organic production and commercialization. The organic market network emphasizes the opening of new markets for organic food, including supermarkets, hotels and restaurants. The ecological farming network highlights the establishment of organic farmers markets throughout the country.

RAE Peru, Grupo Ecologica Peru and ANPE have been grouped in different networks since each one is engaged strongly to a particular set of local and global actors, including other NGOs, producers and international cooperation agencies. However, they converge in several platforms at national, regional and global level. At national level, RAE Peru, Grupo Ecologica Peru and ANPE are part of the Platform 'Peru, Country Free of Transgenics' and the Peruvian Agro-ecological Consortium. Till 2008 RAE Peru and ANPE were members of the National Commission of Organic Products (CONAPO). Together with other NGOs, small scale producers, small-scale agri-enterprises, universities and organic consumers, RAE Peru, Grupo Ecologica Peru and ANPE are part of the Peruvian Agro-ecological Movement (PAM) (Alvarado, 2008, interview; Wu, 2008, interview). PAM is a nation-wide non-formal platform of organizations promoting organic food in Peru (Appendix 2b). At regional level, RAE Peru, the Grupo Ecological Peru and ANPE are associated to the Latin American agro-ecological movement (MAELA). RAE Peru and the Grupo Ecological Peru are also affiliated to the IFOAM Regional Group for Latin America and the Caribbean (GALCI) (Appendix 2a and Appendix 2b) (Alvarado, 2008, interview; Pardo, 2008, interview; Trejo, 2008, interview). At global level, RAE Peru, the Grupo Ecological Peru and ANPE are affiliated to the International Federation of Organic Agriculture Movements (IFOAM). Additionally, ANPE is affiliated to the Slow Food Movement (Appendix 2a and Appendix 2b).

5.2.1 The agro-ecological network

Key actor: the Ecological Agriculture Network of Peru

The Ecological Agriculture Network of Peru (RAE Peru) is the main organization that promotes organic agriculture and the development of organic food enterprises in Peru. It was founded in 1989 and it is the first agro-ecological NGOs founded throughout Latin America (Alvarado, 2008). RAE Peru currently affiliates 16 NGOs widely spread in Peru and it is formally registered under the Peruvian law as a non-profit organization (Table 5.1).

RAE Peru aims to develop organic food enterprises, and to facilitate the development of local organic markets. In order to reach those aims, RAE Peru provides small-scale producers with capital, organizational skills and advice in order to convert their conventional farming practices in organic farming (Alvarado, 2008, interview; Wu, 2008, interview). As key actor of the agro-ecological network in Peru, RAE Peru also provides a set of resources to the network, for instance, experience in network management, network building, communication strategies, facilitation of knowledge towards producer associations, channeling of small grants and lobbing (Alvarado, 2008, interview). In addition to prioritizing local market development and advocacy, RAE Peru gives

Table 5.1. NGOs affiliated to RAE Peru.

Centro de Investigación, Documentación, Educación, Asesoramiento y Servicios (Centro IDEAS)

Instituto de Desarrollo y Medio Ambiente (IDMA)

Asociación para el Desarrollo Rural de Cajamarca (ASPADERUC)

Centro de Apoyo Rural (CEAR)

Asociación Fuerza por la Selva Viva (FUSEVI)

Servicios Educativos Promoción y Apoyo Rural (SEPAR)

Instituto de Ecología y Plantas Medicinales (IEPLAM)

Centro de Investigación, Educación y Desarrollo (CIED)

Asociación Especializada para el Desarrollo Sostenible (AEDES)

Instituto de Medio Ambiente y Género para el Desarrollo (IMAGEN)

Equipo de Desarrollo Agropecuario de Cajamarca (EDAC)

Centro para el Desarrollo Agropecuario (CEDAP)

Red Ecológica Interinstitucional Hatun Sacha (REIHS)

Centro de Estudios Sociales 'Solidaridad' (CES Solidaridad)

Centro de Investigación, Capacitación, Asesoramiento y Promoción (CICAP)

El Taller

Source: Alvarado, 2008, interview; Wu, 2008, interview.

priority to organizational strengthening and capacity building for small-scale organic producers (Alvarado, 2008, interview; Wu, 2008, interview). This last priority is considered a key condition for the organizational sustainability of RAE Peru in the long-term.

RAE Peru operationalizes its aim following five strategies. First, work backed by agro-ecology science; second, generation of successful agro-ecological small-scale enterprises; third, intensive networking with their affiliated and non-affiliated NGOs, fourth, lobbying to mainstream organic production; and finally, raising public awareness of organic products in Peru (Alvarado, 2008, interview).

Democratically elected representatives of affiliated NGOs and individuals constitute the national coordination board of RAE Peru. The national coordination board has three main responsibilities: (1) coordinating the implementation of the strategic plan of RAE Peru; (2) organizing the National Meeting of Agro-ecological Agriculture each two years where all affiliates establish a master agenda to guide action of the network; and (3) supporting their affiliated NGOs logistically and cofinancially (Alvarado, 2008, interview). In the most recent election of new board members (September 2010) a new national coordination board for the period 2010-2012 was elected. A representative of the Institute of Development and Environment (IDMA) was elected president and an individual associate the vice-president. The rest of representatives belong to El Taller, CIED, Centro IDEAS, CICAP, SEPAR, FUSEVI and another individual associate (RAE Peru website, 2012).

According to Alvarado (2002), RAE Peru evolution has unfolded in three periods: (1) scientific foundation and capacity building from 1989 to 1994; (2) organizing organic producers from 1995 to 2000; and (3) lobby and market development from 2001 onwards.

Network relationships: cooperation

Since its foundation, RAE Peru has worked in cooperation with its affiliated NGOs, organic producers, universities, governmental agencies and international cooperation agencies to promote organic agriculture among small scale producers and consumers. RAE Peru works most closely with one of its affiliated NGOs; the Center for Research, Documentation, Education, Advice and Services (Centro IDEAS). Centro IDEAS has been head of the board for several periods and its role has been central in most of the achievements of RAE Peru (Wu, 2008, interview; Alvarado, 2008, interview).

RAE Peru also collaborates closely with the network of alternative agriculture (RAAA), ANPE, the Committee of Ecological Consumers (CCE) and other agro-ecological NGOs (Alvarado, 2008, interview; Wu, 2008, interview). They implemented jointly projects and shared responsibilities in common platforms. For instance, RAE Peru joined efforts with RAAA to organize the organic producers throughout Peru and supported the formation of ANPE as a representative organization of all organic producers. Coordination of RAE Peru and their affiliated NGOs is done through national meetings where the agenda for cooperation is set. The whole Peruvian agro-ecological movement and the organic producers have both their own national meetings each two years. In September 2008, the Peruvian agro-ecological movement had its XIth National Meeting of Ecological Agriculture (XI ENAE) (Alvarado, 2009, interview). For more details of the Peruvian agro-ecological movement (PAM) see Appendix 2b.

Since its foundation RAE Peru has played a leading position in promoting ecological agriculture and developing organic markets in Peru. For instance, in 1989, NGOs affiliated to RAE Peru implemented the 'Biocanasta', an initiative to deliver organic fresh food from farmers directly to consumers. Centro IDEAS and the Institute of Development and Environment (IDMA), NGOs affiliated to RAE Peru, played a key role in setting up Grupo Ecologica Peru and the farmers' market Bioferia in the district of Miraflores in Lima. RAE Peru has also set up the first organic certifier body in Peru, Inka-Cert, in 1994 (since 1998 it changed its name to Bio Latina operating in Peru, Nicaragua, Colombia and Bolivia), ANPE, the Committee of Ecological Consumers (CCE; founded in 2002), the Bio-store K'ANTU, and the farmers' market Bioferia in the district of San Borja in Lima (together with Grupo Ecologica Peru). Centro IDEAS and Taller have also implemented a Bioferia in Arequipa, a region in the south west of Peru (Alvarado, 2002; Alvarado, 2008, interview; Wu, 2008, interview). In order to raise public awareness about organic food consumption and healthy life-style, in 2006 RAE Peru, Red Peru and CCE started the radio program 'Compartiendo Radio' at San Borja radio in Lima. During the last three years 116 programs have been produced. Since 2009 the program is also broadcasted at internet (Alvarado, 2008; Alvarado, 2008, interview; Alvarado, 2009, interview; Wu, 2008, interview).

RAE Peru collaborates not only with organizations working with organic products but also with 'conventional' producers. For instance, RAE Peru collaborates with the Peruvian Association of Consumers and Users (ASPEC), the Economy of Solidarity Network in Peru (GRESP) and the

National Union for Coffee Producers (JNC) (Alvarado, 2008, interview; Wu, 2008, interview) (Appendix 2a and Appendix 2b). RAE Peru also works closely with the Consortium of Private Organizations for the Promotion of Small and Medium Business Development of Peru (COPEME). The Centro IDEAS in partnership with COPEME is providing trainings to producers in market instruments, supply strategies, and demand driven strategies (Alvarado, 2008, interview).

RAE Peru also participates very actively in several national platforms working with small scale producers beyond the agro-ecological NGOs, including the National Convention for Peruvian Agriculture (CONVEAGRO), the National Agrarian Confederation (CNA), the National Union for Coffee Producers (JNC), the Program for Exchange, Dialogue and Consultation on Sustainable Agriculture and Food Sovereignty (PIDAASSA PERU) and the Coordinadora Rural. RAE Peru also belongs to the Peruvian Network of Fair Trade and Ethical Consumption and the National Environmental Society (SNA). The former platform is an association of small enterprises, not only food producers but also shoe makers, garment and craft makers. The latter platform represents a large number of agro-ecological and agrarian organizations (Alvarado, 2008, interview; Trejo, 2008, interview; Wu, 2010, interview) (Appendix 2b). In 2010, RAE Peru established cooperative ties with the Peruvian Society of Gastronomy (APEGA). Since 2008 APEGA organizes annually the Peruvian food festival 'Mistura' (Wu, 2010, interview). Small scale producers, especially organic producers, provide organic products to the festival (APEGA website, 2010). In 2010, 40% of all shops of Mistura offered ecological food (Alvarado, 2010, interview)

In 2001, together with ANPE and the National Agrarian University of Peru (UNALM), RAE Peru has played a key role in the formation of the National Commission of Organic Products (CONAPO) by the Peruvian government (Appendix 2b). Till 2008 RAE Peru belonged to CONAPO as well (Appendix 2a,b) (Alvarado, 2008, interview; Trejo, 2008, interview). RAE Peru has contributed to the CONAPO's organizational structuring, the National Plan of Strengthening the Organic Agriculture and the national regulation development in organic production (Alvarado, 2002; Alvarado, 2008, interview). In 2008 the National Commission of Organic Products became the National Council of Organic Products (keeping the acronym CONAPO). The National Commission of Organic Products was dissolved as its aim was reached: the elaboration of a proposal for the national promotion of organic agriculture and its approval in 2008 as the law N°29196 'Promotion of Organic Agriculture'. However, up to now (2012) the National Council of Organic Products (the new CONAPO) is not operational yet (Compartiendo N° 10 and N° 11, 2012).

Inputs of NGOs have been key to the development of small scale producers and small scale enterprises, but not to the medium and large scale organic producers. During the last 25 years, agro-ecological NGOs laid the foundations for the organic production oriented to the market in Peru working together with platforms, associations, unions and cooperatives of small scale organic producers. However, the inputs of NGOs towards medium and large scale organic producers have been scarce. As larger producer organizations have substantially more resources, they usually do not need support from NGOs. They get (financial) support for developing their businesses from banks and investors. For instance, at the beginning COCLA, a Peruvian large coffee cooperative that supplies coffee to CRAFT general foods, received support from this large buyer. CRAFT is one of the larger global buyers of coffee representing 20.8% of the global market. COCLA used to produce and sell only conventional coffee. CRAFT organized producers to supply organic coffee for a demanding international market of organic coffee. In this case, the market demand and the big buyer CRAFT triggered the conversion to organic production. Agro-ecological NGOs did not have any role at COCLA. Similarly, JNC, affiliating mostly larger coffee enterprises, does not have an interest in investing to build local market for coffee. According to the head of RAE Peru (Alvarado, 2008, interview), 'larger scale producers are not interested in developing small scale producers'.

RAE Peru has limited ties and little cooperation with Peruvian governmental agencies. Identified connections are with the National Service of Agrarian Health (SENASA), the National Institute for Agrarian Innovation (INIA), the Cooperation Fund for Social Development (FONCODES) and the Ministry of Agriculture's General Direction of Agrarian Competitiveness (DGCA) (Appendix 2e) (Alvarado, 2008, interview; Wu, 2008, interview). Cooperation of RAE Peru with governmental agencies rests on personal motivation of governmental officials, not on institutional policies of governmental agencies (Alvarado, 2008, interview; Wu, 2008, interview; Pardo, 2008, interview; Trejo, 2008, interview). Moreover, governmental agencies usually have their own agenda. For instance, the National Service of Agrarian Health (SENASA) focuses on large scale producers for organic export and international markets, not on small-scale organic producers and local markets. Little research is done on ecological production and commercialization at the national agricultural research centers SENASA, the National Institute for Agrarian Innovation (INIA) and the National Agrarian University (UNALM) because of limited budget and experts. During the last years an institutional infrastructure is being created to support organic production and develop organic market, but it is far from fulfilling expectations (Alvarado & Wu, 2008, interview; Trejo, 2008, interview). In the strategy of NGOs to influence the Peruvian government, consumers are becoming a key ally for NGOs.

At the international level, RAE Peru is connected with several regional and global platforms of organic production. For instance, RAE Peru is part of the Latin American Agro-ecological Movement (MAELA), the IFOAM Regional Group for Latin America and the Caribbean (GALCI) and the International Federation of Organic Agriculture Movements (IFOAM) (Wu, 2010, interview). RAE Peru is also connected with international cooperation agencies of global reach. They provide policy guidance, technical support and funding to RAE Peru. RAE Peru has long-term cooperative ties with the Dutch interchurch organization for development cooperation (ICCO). ICCO has provided small grants to RAE Peru from 2000 till 2011 for projects aimed at institutional strengthening, market development and access for small scale organic producers, capacity building and policy influence (Wu, 2010, interview). Currently (2012) ICCO collaborates with RAE Peru as part of a long-standing collaboration with the Agro-ecological Consortium (ICCO-KIA website, 2012). RAE Peru has also received support of other international cooperation agencies: Heifer International from 2006 to 2008, the Food and Agriculture Organization of the United Nations (FAO) from 2004 to 2006 and The German catholic bishops' organization for development cooperation (MISEREOR) from 1989 to 1996. Before 2000, the German Development Service (DED), Bread for the World and MISERIOR supported RAE Peru in implementing capacity building projects (Alvarado, 2008, interview; Wu, 2008, interview).

Network relationships: power

This section analyzes the power relations in the agro-ecological network, focusing on the key organizations identified in Peru and their international connections. The national coordination board of RAE Peru is the national node of the network. Local nodes are the local NGO constituencies, including the Center for Research, Documentation, Education, Advice and Services (Centro IDEAS). The above mentioned regional and global partners are key actors at the supra-national level.

Conflicts are present in the agro-ecological network in Peru and occur mainly between the coordination boards of national NGOs. Specifically, conflicts occur between the RAE Peru national coordination board (Centro IDEAS has the central position), the ANPE national coordination board and the Grupo Ecologica Peru's coordination board (Alvarado, 2008, interview; Pardo, 2008, interview; Trejo, 2008, interview; Wu, 2008, interview; Wu, 2010, interview). For a long time, the NGOs of RAE Peru controlled Grupo Ecologica Peru and the Bioferia Miraflores, considering them the best proofs of their development achievements.

RAE Peru complains that ANPE is not doing what it should do according the task division agreed upon in the Peruvian agro-ecological movement (PAM) (Alvarado, 2008, interview; Wu, 2008, interview). However, the basic tension between RAE Peru and ANPE has funding reasons (Trejo, 2008, interview). At the beginning ANPE depended on RAE Peru financially. But, over time ANPE gained more independency by centralizing decision making and increased funding (Trejo, 2008, interview). Now, local associations of producers affiliated to ANPE have to wait for ANPE's national coordination board decision for action. Thus, NGOs affiliated to RAE Peru and associations of producers affiliated to ANPE cannot coordinate between them directly at the local level. This has become an obstacle for RAE Peru (Alvarado, 2008, interview). In several regions of Peru there are power struggles among ANPE's national board and local NGOs affiliated to RAE Peru. This happened between the NGO CEAR and the producer associations of Huancayo. In Piura and Arequipa similar developments occurred.

At the ground level, there is not much struggle between ANPE's affiliated producers and RAE Peru's affiliated NGOs (Alvarado, 2008, interview; Wu, 2008, interview). According to Silvia Wu (2010, interview), former director of RAE Peru, local producer associations affiliated to ANPE are glad to work with NGOs and they value the support of RAE Peru: 'The problem is with the ANPE's national coordination board but not with their affiliated producers'. ANPE and the Grupo Ecological Peru are seen by RAE Peru's NGOs as 'immature' organizations (Alvarado, 2008, interview; Wu, 2008, interview). This struggle has made that RAE Peru stimulates the formation of new local associations of producers outside of ANPE, for example in the case of the Bioferia of Arequipa (Alvarado, 2008, interview).

Conflicts also occur between NGOs and producers. Grupo Ecologica Peru has become the main 'battle field' between agro-ecological NGOs and organic producers. Conflicts started in 2004 when a producer representative became president of the organization. Before that year, only NGO members used to lead the organization, not producers (Pardo, 2008, interview). With a better position of producers in the coordination board of Grupo Ecologica Peru, the affiliated NGOs, particularly Centro IDEAS, IDMA, CEAR, Huayuna and CANDELA Peru have lost position. For this reason, NGOs, especially Centro IDEAS, look at the Grupo Ecologica Peru with mistrust and see it as a competitor for funding and power. They criticize the way Grupo Ecologica Peru manages the Bioferia Miraflores and its isolation of the Peruvian agro-ecological movement (PAM) (Alvarado, 2008, interview). For this reason, recently Centro IDEAS has set up another Bioferia in the district of San Borja in Lima. Centro IDEAS provides only logistical resources and advice but not funding (Alvarado, 2008, interview). During 2010 the conflict has become stronger. In the last change of board members the Centro IDEAS become the head of the board of Grupo Ecologica Peru. The new coordination board of Grupo Ecologica Peru, established for the period 2009-2011, had the task to harmonize producers and NGOs interests. In contrast to the former coordination board led by producers, the current board led by NGOs aims to reduce the conflicts between producer and NGO representatives (Alvarado, 2010, interview). However, rather than diminishing conflicts, the election of a NGO representative as president has enhanced conflicts. As result, in January 2011 the NGOs Centro IDEAS, IDMA and CEAR affiliated to RAE Peru, joined by a group of producers, almost left Grupo Ecologica Peru and a new coordination board of Grupo Ecologica Peru almost was appointed.

International NGOs see conflicts evolving between local NGOs and producers as a risk to their aid policy aims. For instance, the success of the actual aid policy based on market access depends on cooperation of actors. Disputes and conflicts among local actors erode further cooperation. However, the increasing attention of international NGOs towards small producers is directly contributing to deeper power struggles among RAE Peru, Grupo Ecologica Peru and ANPE. While the number of international NGOs providing funding to RAE Peru is decreasing, the number of international NGOs interested in funding Grupo Ecologica Peru and ANPE is increasing (Alvarado, 2008, interview; Pardo, 2008, interview).

5.2.2 The organic market network

Key actor: Grupo Ecologica Peru

Grupo Ecologica Peru, founded in 1998, is a networked NGO integrating 5 NGOs and 24 organic producers, including associations and individual producers (Table 5.2). Before 2001 the organization's former name was just 'Ecologica Peru'. The founders of Grupo Ecologica Peru are the NGOs IDMA, Huayuna and Centro IDEAS. Later the NGOs Candela Peru and CEAR, and a pool of organic producers have been incorporated.

Table 5.2. NGOs and SMEs affiliated to Grupo Ecologica Peru.

5 NGOs (Centro IDEAS, IDMA, Huayuna, CEAR and CANDELA Peru)

15 organic producer associations, including BioFrut, Monticielo, ecological producer association Valle Santa Cruz-Satipo, Valle Sano, producer association of Maca (APROMAC) and the ecological producer association of San Jerónimo of Surco (APES)

9 individual organic producers, including BioAgricultura Casa Blanca, BioHuertos del Manantial, BioGranjas, La Cabrita, Vacas Felices

Source: Pardo, 2008, interview.

The aim of Grupo Ecologica Peru is the development of local organic markets and the commercialization of organic products supplied by small scale producers. Since its foundation, Grupo Ecologica Peru has led, in cooperation with NGOs affiliated to RAE Peru, the Bioferia of Miraflores in Lima to commercialize organic products directly to consumers. Moreover, Grupo Ecologica Peru organizes the supply of organic products to supermarkets. Management of market-oriented production and commercialization of organic food are new and challenging roles for Grupo Ecologica Peru (Pardo, 2008, interview). It is worth mentioning that the Bioferia Miraflores and the supply to supermarkets are self-financed projects of Grupo Ecologica Peru. None of these two projects is financed by international NGOs. International NGOs are mainly financing the upgrading and replication of those projects in Peru and neighboring countries of the region (Pardo, 2008, interview).

Grupo Ecologica Peru has moved from a 'volunteer' organization to a professional, structured organization. In 2006 a formal organizational structure was adopted, including the following aspects: accounting, administrative staff and standardized procedures and norms within the organization and within the Bioferia Miraflores (Pardo, 2008, interview).

In order to move forward as an independent organization, Grupo Ecologica Peru is currently redefining its organizational structure, for instance, the organizational membership. According to the former coordination board for the years 2007-2009, a priority was to define who is member and who is not. Regarding the issue of membership, Grupo Ecologica Peru has 29 affiliated members but, for instance, only 12 are paying their fees. The conflicts among affiliated organizations have been reason to postpone defining this situation (Alvarado, 2010, interview).

Network relationships: cooperation

First of all, Grupo Ecologica Peru has cooperative ties with their providers of organic food. The providers consist of about 26 small scale organic food producers located throughout Peru (Alvarado, 2010, interview).

Huayuna is the key partner of Grupo Ecologica Peru. Huayuna is an NGO specialized in ecological farming. Huayuna provides knowledge in ecological farming practices and agro-ecological technology to producers providing to Grupo Ecologica Peru. Grupo Ecologica Peru also works closely with RAE Peru, the International Center of Potato (CIP), the Peruvian Association of Consumers and Users (ASPEC), the Municipality of Miraflores, the Municipality of San Isidro and the Committee of Ecological Consumers (CCE) to strength the Bioferia of Miraflores (Pardo, 2008, interview). However, different from most agro-ecological NGOs, Grupo Ecologica Peru has been quite isolated from the activities of RAE Peru, RAAA and ANPE (Alvarado, 2009, interview).

Grupo Ecologica Peru approached supermarkets and other potential channels of commercialization of organic products after noticing that the Bioferia Miraflores had limits in organic business growth. Supermarket owners got interested and requested a formal commercial relation with Grupo Ecologica Peru. As a legal status was a requirement for supermarkets, Grupo Ecologica Peru got a legal status in 2001 establishing commercial ties with local supermarkets of Lima (Pardo, 2008, interview). Grupo Ecologica Peru supplies organic food to 20% of all supermarket stores located in Lima, including the supermarkets Plaza VEA, Santa Isabel, Vivanda and Tottus (Alvarado, 2010, interview; Pardo, 2008, interview). Santa Isabel was the first supermarket interested in buying organic products. According to Pardo (2008, interview) influences from Europe on market tendencies of organic food and organic shops have been helpful to access supermarkets in Peru. However, the volume of supply is the major bottleneck of small scale producers to access supermarkets. For this reason, the priority of Grupo Ecologica Peru is to support small scale producers to improve their production in order to increase supply.

Most food suppliers of supermarkets are medium and large sized companies. Grupo Ecologica Peru is the first supplier of organic food to supermarkets and the largest supplier of organic food that is an NGO. Currently, there are two companies Santa Natura and Hidroplant S.A that provide organic food to supermarkets. It is important to point out that in the last years also a few small scale organic producers supply directly to supermarkets, including the producers of the town of Cieneguilla under the brand Max banana. This shows that the demand of organic food by supermarkets is not fully covered yet (Alvarado, 2010, interview; Pardo, 2008, interview).

Like RAE Peru (see above) Grupo Ecologica Peru has established cooperative ties with the Peruvian Society of Gastronomy (APEGA) and the Peruvian Food Fair 'Mistura' (Alvarado, 2010, interview). According to Alvarado (2010, interview) the only way to respond to the increasing demand of organic products is organizing supply. Grupo Ecologica Peru has a leading role in developing the organic market by transferring its experience to the Bioferias established throughout Peru. As Alvarado stated (2010, interview) 'by strengthening the Bioferias, Grupo Ecologica Peru gets stronger. By working in cooperation, Grupo Ecologica Peru and the Bioferias get stronger; for instance, to jointly lobby against municipalities that are not eager to support the setting up of new Bioferias'. Additionally, Grupo Ecologica Peru cooperates with restaurants and catering companies to supply them organic food.

Consumers have contributed to the growth of the Bioferia Miraflores. Consumers provide suggestions and feedback to producers, for instance, in terms of diversification and volume of products and to Grupo Ecologica Peru, for instance, on how to better organize the Bioferia and on options of further commercialization. Cooperation in terms of feedback between producers and consumers has been central in the development of the Bioferia Miraflores.

Grupo Ecologica Peru has connections with international platforms and international cooperation agencies. In developing local organic markets, Grupo Ecologica Peru works closely with the International Federation of Organic Agriculture Movements (IFOAM) and the Bolivian NGOs AgroecolAndes, Tierra Viva and AOPEP (Pardo, 2008, interview). The collaborating international cooperation agencies are: the Dutch Catholic Organization for Relief and Development Aid (CORDAID), Dutch OXFAM-NOVIB, the the Dutch interchurch organization for development cooperation (ICCO) and the German Schmitz Foundation. CORDAID cooperates with Grupo Ecologica Peru in issues related to local market access for organic products coming from small scale producers. In particular, the support of CORDAID is intended to transfer and extend the knowledge of Grupo Ecologica Peru towards local partners of CORDAID, El Taller, Imagen, Sicab and Huayuna (Pardo, 2008, interview). CORDAID is currently providing funding (2009-2012) to implement the 'National Network of Bioferias' of Peru and strengthen the commercial capacity of the Bioferias. As a part of this task, Grupo Ecologica Peru organized the 1st congress of Peruvian Bioferias in 2010.

OXFAM-NOVIB provides financial support to cooperate with Grupo Ecologica Peru to organize the supply of organic products to local markets. The last financial support extended from

2007 to 2010. The experience of Biofrut, a producer association that produces fruits and vegetables, has been key to get support from OXFAM-NOVIB. The support aimed to develop institutional capacity and improve production, for instance, of the processing plant, and provide technical assistance for providers. In addition, during that period, OXFAM-NOVIB has collaborated with Grupo Ecologica Peru to improve its production and commercialization capacities, for instance by introducing Manufacturing Good Practices (MGP) and ecological certification. During this period, Grupo Ecologica Peru has been granted with the BSS certification to process and certify organic products. BSS is a German certifier. OXFAM-NOVIB will not provide financial support to Grupo Ecologica Peru anymore for the future due to the withdraw from Latin America, nevertheless it is providing contacts with other potential sources of funding (OXFAM-NOVIB, 2011a,b).

ICCO is collaborating with Grupo Ecologica Peru by providing funding to the Agro-ecological Consortium. The aim of this two years cooperation is to influence public policies on local commercialization of organic food and Participatory Guaranty Systems (PGS), and build support from local authorities for the Bioferias. During the former coordination board, Grupo Ecologica Peru got financial support from the German Schmitz Foundation to implement Hazard Analysis and Critical Control Points (HACCP) and Agricultural Good Practices for its small scale providers of organic food (Alvarado, 2010, interview; Pardo, 2008, interview).

Network relationships: power

Grupo Ecologica Peru is the national node in the network. The network has sub-national nodes of NGOs and producers. At the global level, key actors are the Dutch Catholic Organization for Relief and Development Aid (CORDAID) and Dutch OXFAM-NOVIB, as explained above. The global platform is the International Federation of Organic Agriculture Movements (IFOAM). Key allies at regional level are the Bolivian NGOs AgroecolAndes, Tierra Viva and Association of ecological producer organizations of Bolivia (AOPEP).

NGO and producer members of Grupo Ecologica Peru are involved in power struggles. Producers put pressure on NGOs and challenge their leadership within Grupo Ecologica Peru. In 2004 producers took over the organizational structure of Grupo Ecologica Peru. NGOs were commissioned to take the role of 'advisors' of Grupo Ecologica Peru's board of directors (Pardo, 2008, interview). However, in the elections of the board members of Grupo Ecologica Peru in 2009 NGOs regained power positions. The NGOs Centro IDEAS and CEAR took power together with three organic enterprises: the Campiña, the Cabrita and Korin. This furthered the conflicts between NGOs and producers, resulting in the almost resign of the NGO members of the board and other NGO members of RAE Peru (Alvarado, 2009, interview; Alvarado, 2010, interview). This power struggle between NGOs and producers had also impacted on the conflict in the RAE board (see above). The struggle between NGOs and producers affiliated to Grupo Ecologica Peru has to do with gaining power in the organization, controlling its resources and the Bioferia Miraflores, and getting legitimacy from international funding agencies.

Power struggles occur also between producers themselves. Producers involved in each power struggle are affiliated to Grupo Ecologica Peru, the Bioferia Miraflores and ANPE. The Bioferia Miraflores intends to be an independent organization from Grupo Ecologica Peru and seeks financial autonomy, but Grupo Ecologica Peru wants to keep the Bioferia Miraflores under its control. According to Pardo, the former head of Grupo Ecologica Peru (2008, interview) 'most producers that are part of the Bioferia Miraflores do not care much about concerns that Grupo Ecologica Peru raises. Most producers are not interested in the organizational development of Grupo Ecologica Peru. They just want to commercialize their products in the Bioferia since it is a good business opportunity for them'. The board members claim that producers of Bioferia Miraflores only use their power for 'self-benefit' and not for the organization. Bioferia producers claim that they pay fees to Grupo Ecologica Peru for commercializing in the Bioferia Miraflores and they do not know what is done with the collected money. The board members claim that the fees are necessary and they are not enough to cover operational costs. The Bioferia Miraflores has a turn-over of 1 million soles/year. Moreover, the fact that the Grupo Ecologica Peru's board was directly financing the supply to supermarkets brought mistrust from the Bioferia Miraflores' affiliated producers. They claim that Grupo Ecologica Peru was possibly subsidizing its supermarket supply initiative with the Bioferia Miraflores' revenues (Alvarado, 2008, interview; Wu, 2008, interview). According to Pardo (2008, interview) much of the conflict between Grupo Ecologica Peru and the Bioferia Miraflores has to do with the transition from subsidized to self-financed projects. She also claims that producers do not want to pay the price of economic growing of the Bioferia Miraflores. Additionally, conflicts among producers also involve producers affiliated to ANPE. Grupo Ecologica Peru and ANPE have different views and approaches of how producers should participate in the market. They use different schemes to guarantee the organic condition of food products. While Grupo Ecologica Peru only admits producers with organic certification to commercialize organic products in the Bioferia Miraflores, ANPE promotes Participatory Guaranty Systems (PGS) as organic guaranty for small scale producers. Grupo Ecologica Peru claims that ANPE is not really working enough to developing organic market in Peru and ANPE claims that economic motivations rather than social motivations are the main drivers for Grupo Ecologica Peru to promote organic markets (Alvarado, 2008, interview; Trejo, 2008, interview).

5.2.3 The ecological farming network

Key actor: the National Ecological Producers Association of Peru

The National Ecological Producers Association (ANPE), founded in 1998, is a networked NGO of small scale organic producer associations, including small food processors and family small-scale enterprises. ANPE is the largest organization of organic producers in Peru. In total about 15000 organic producers are affiliated to ANPE, organized in 22 producer associations (Table 5.3). The affiliated producer associations are mainly oriented to producing and commercializing organic food in local markets all over the country. The more active affiliated organizations are the producers of Alto Piura (APPEAP), the producer association of Satipo, the producer association of Ucayali, the producer association of Puno and the producer association 'Biofrut' of Mala-Lima. Besides ANPE, this producer association is also affiliated to Grupo Ecologica Peru.

ANPE works on promoting organic farming among small scale producers and facilitating the development and access of small enterprises to the market. ANPE is building up the organic market in Peru. While the Bioferia Miraflores was mainly promoted by Grupo Ecologica Peru, ANPE has established 13 other Bioferias in several regions of Peru, including Huanuco, Huancayo,

Table 5.3. Organic producer associations affiliated to ANPE.

---

20 local organizations of producers and 2 local women organizations are affiliated to ANPE. It means
   12,000 organic producers organized in 22 sub-national regions throughout Peru.
Associations of organic producers affiliated to ANPE:
   Asociación de Productoras(res) Ecológicas del Alto Piura (APPEAP)
   Asociación Regional de Productores Ecológicos de Lambayeque (ARPEL)
   Asociación Regional de Productores Ecológicos de La Libertad (ARPELL)
   Asociación Regional de Productores Ecológicos de Cajamarca (APERC)
   Asociación Comunal de Productores Agroecológicos (ASCOPAE)
   Asociación Regional de Productores Ecológicos de Ancash (ANCASH)
   Asociación Regional de Productores Ecológicos de Lima (ARPEL)
   Asociación Departamental de Productores Ecológicos – Huanuco (ADPEH)
   Asociación de Productores Ecológicos de la Región Centro (APEREC)
   Asociación Regional de Productores Ecológicos de Huancavelica (ARPEH)
   Asociación Regional de Productores Orgánicos de Ayacucho-Huamanga (ARPOA)
   Asociación Regional de Productores Ecológicos Apurimac (ARPEA)
   Asociación Regional de Organizaciones, Transformadores y Productores Ecológicos del Cusco
      (ARPEC)
   Asociación Regional de Wiñay Warmis Cusco (Wiñay Warmis)
   Asociación Nacional de Productores Ecológicos del Perú Ayaviri (ANPE-AYAVIRI)
   Asociación Regional de Productores Ecológicos de Arequipa (ARPEA)
   Comité de Productores Agroecológicos del Valle de Moquegua (COPAEM)
   APROCEP de Tacna
   Asociación Regional de Productores Ecológicos de San Martín (ARPESAM)
   Federación de Productores Ecológicos de Ucayali (FEPEU)
   Asociación de Conservacionistas de la Agro-biodiversidad de la Provincia de Ucayali – Loreto
   Asociación de Mujeres Agro-Ecologistas de la región de Ucayali (AMARU)

Source: Trejo, 2008, interview; Trejo, 2009, interview.

Lambayeque, Cajamarca, Cusco, Puno, Piura, Junín, Ayacucho, Apurimac, Arequipa, Pucalpa and
Lima (the Bioferia Cieneguilla). Additionally, ANPE devotes efforts to lobbying and advocating
for adequate policy infrastructure on organic production and fair market conditions for organic
producers in Peru (Trejo, 2008, interview; Trejo, 2009, interview).

   ANPE's producers guarantee the organic properties of the products commercialized at the 13
Bioferias by the adoption of the Participatory Guarantee System (PGS) approach. Most organic
producers commercializing their products at local markets in Peru have PGS, and not a third
party certification. Only producers that commercialize organic food in the Bioferia of Miraflores
have certification from a third party certification body.

   Since 1986 ANPE organizes the National Meeting of Ecological Producers (ENPEs) every two
years. The first ENPE was organized in commission by the agro-ecological NGOs RAE Peru and

Centro IDEAS. ENPE XIII was organized in 2009 and ENPE XIV was organized in 2011 (Trejo, 2010, interview). ANPE has brought new dynamics to the Peruvian agro-ecological movement (PAM), usually overwhelmed by conventional NGOs. Besides building an organizational framework for organic producers, ANPE has contributed to the establishment of Grupo Ecologica Peru (Trejo, 2008, interview).

Network relationships: cooperation

Since its foundation ANPE has worked in cooperation with its affiliated producers, NGOs, universities, governmental agencies and international cooperation agencies to promote organic agriculture among small scale producers and consumers. The contribution of producers affiliated to ANPE is key in terms of land, labor and knowledge on traditional farming (Trejo, 2008, interview).

At the national level, the national coordination office of ANPE works in close cooperation with their affiliated regional producer associations and other actors, including governmental agencies, sub-regional governments, municipalities, local NGOs and producer unions, to support small scale producers. For instance, ANPE works closely with the ecological producer association of aromatic plants of Sihuas (APEPA). APEPA is the regional branch of ANPE in the region of Ancash and it is currently exporting aromatic herbs to foreign markets. The achievements of APEPA have triggered the formation of other organic producer associations in several provinces of Ancash, including Ayja, Carhuaz, Recuay and Huari. In cooperation with ANPE these associations have set up a local base of organic producers in Ancash. Similar examples exist in other sub-national regions of Peru (for a complete list, see Table 5.3).

ANPE has ties with other agro-ecological NGOs. ANPE works closely with RAE Peru and RAAA. They provide ANPE with trainings, knowledge and lobby expertise. ANPE connects them to producers. ANPE works closely with NGOs affiliated to RAE Peru, especially IDMA in the region of Huánuco, CES Solidaridad in the region of Lambayeque, Centro IDEAS in the region of Piura and Arariwa in the region of Cusco. These NGOs provide to ANPE's affiliated producers advice on organic farming. In reaching consumers, ANPE collaborates with the Peruvian Association of consumers and users (ASPEC) to organize public campaigns in favor of organic food (Trejo, 2008, interview). It is important to point out that agro-ecological NGOs cooperate on a temporary or a continuous base. An example of the former is the joining of RAE Peru, ANPE and CCE to elaborate the proposal of the national regulation for organic production in Peru. An example of the latter is the setting up of the Platform 'Peru, country free of transgenics', the National Commission of Organic Products (CONAPO) and the Peruvian Agro-ecological Consortium by the aforementioned three NGOs.

ANPE also has ties with several organic and fair trade associations of small scale producers that are not their constituencies. They are: CEPICAFE, the JNC – committee organic coffee growers, the network of organic banana producers (REPEBAN), the association of small-scale producers of organic bananas from Piura (CEPIBO) and the association of producers of mango of Alto Piura (APROMALPI). CEPICAFE, APROMALPI and CEPIBO supply organic products to local and international markets (Trejo, 2009, interview). Recently these associations and the associations affiliated to ANPE are joining in platforms. For instance, CEPICAFE, APROMALPI, CEPIBO and other associations located in the northern part of Peru have joined ARPEP. Larger scale organic producers are not affiliated to ANPE. They are affiliated to other organizations, such as COCLA and JNC, international cooperation agencies and exporters (Pardo, 2008, interview). Similar to RAE Peru and Grupo Ecologica Peru, ANPE has also built up cooperation ties with the Peruvian Society of Gastronomy (APEGA). Especially, ANPE and Grupo Ecologica Peru channel the supply of organic food products from their organic producer constituencies to the food fair Mistura. This cooperation with APEGA, named 'farmer-cooker' alliance', is opening new markets for small scale producers including top international restaurants and hotels in Peru.

ANPE has also established cooperative ties with non-organic producer associations and agrarian unions. They include among others the National Agrarian Confederation (CNA), the National Convention for Peruvian Agriculture (CONVEAGRO), the Peasant Confederation of Peru (CCP) and the National Users Council of the Irrigation Districts of Peru (JNUDRP) (Appendix 2d). Till 2009 ANPE, CNA and CCP used to work separately despite the fact that all of them represent small scale producers. For instance, while CNA was trying to get a national policy to implement the peasant markets, CCP and CONVEAGRO aimed to reduce the debt of small scale producers (the so called 'PREDA' law). At the meantime, ANPE was trying to establish a regulation for the national organic production. In 2010 the three organizations signed an agreement to work together (Trejo, 2010, interview). The main factor bringing these organizations together was the need to influence collectively the national governmental policies since their isolated actions failed to do so. Cooperation between the three organizations not only takes place at the level of the coordination boards but also at the level of the technical staff. Reaching international markets is also motivating these actors to work together. For instance, CNA, CCP, ANPE, the National Union for Coffee Producers (JNC) and the Peruvian association of wheat and barley (APETRICES) are coming together to coordinate the supply of organic products for international markets. Another issue bringing together ANPE with non-organic producer associations and agrarian unions is the awareness at policy level of the importance of small scale producers for the national economy and the mitigation of climate change. CNA, CCP, ANPE, JNC and APETRICES are collaborating to face the impact of climate change in agriculture.

ANPE has links with national and sub-national platforms promoting organic agriculture in Peru. ANPE is part of the 'Peru Country Free of Transgenics' and the Agro-ecological Consortium. The Peru Country Free of Transgenics has contributed strongly in the approval of the national law to label transgenic foods. Currently the National Congress is discussing and elaborating this law. This platform has also contributed to postpone the entering of transgenic food and seeds in Peru during the next 10 years. According to Trejo, 'big companies have a stake in and will pressure to change this decision'. The Agro-ecological Consortium is an umbrella platform of agro-ecological NGOs and organic producers established in 2009. It was founded by ANPE, RAE Peru, RAAA and the Peruvian Association of consumers and users (ASPEC) to encourage the cooperation of NGOs and producers and reduce overlapping and conflicts between their representatives in order to negotiate a better position with the national government (Alvarado, 2009, interview; Trejo, 2009, interview). Under the coordination of ANPE, currently the Agro-ecological Consortium is working on the development of local organic markets and the development of the Participatory Guaranty Systems (PGS). Members of the platform are elaborating proposals for sub-national regional governments and municipalities, and influence the national government and the local agrarian offices to recognize PGS as a feasible commercialization strategy at local markets (Alvarado, 2008, interview; Trejo, 2008, interview). For instance, in the sub-national region of Huanuco, PGS has been recognized as a valid strategy of commercialization. The idea is to extend this example to other sub-national regions of Peru. This implies coordination with the sub-national regional bureau, municipalities, the National Service of Agrarian Health (SENASA) and the Ministry of agriculture. This project is being implemented for two years in the sub-national regions of Cusco, Huanuco, Lambayeque and Lima (Trejo, 2010, interview). Additionally, together with IDMA, ANPE is part of the Program for Exchange, Dialogue and Consultation on Sustainable Agriculture and Food Sovereignty (PIDAASSA PERU), a platform that represent NGOs promoting sustainable agriculture (Appendix 2d).

ANPE has ties with governmental agencies. ANPE has ties with the Ministry of Agriculture's General Directorate of Agrarian Competitiveness. ANPE also collaborates with the Presidency of the Ministries' Council (PCM), a governmental agency that is promoting the national 'Buy from Peru' campaign with the aim to promote the consumption of locally produced products (Trejo, 2008, interview; Trejo, 2009, interview). However, the main target of ANPE is not anymore the national government but sub-national regional bureaus and municipalities. 'The Ministry of Agriculture is just a paper work office. The money is in the sub-national regional bureaus and municipalities. Now, political and economic decision making depends on sub-national regional bureaus (…). The strategy is to build up cooperation ties with sub-national regional bureaus and municipalities' (Trejo, 2010, interview).

At international level, ANPE has ties with producer associations, international platforms and international cooperation agencies. ANPE works closely with the Association of Ecological Producer Organizations of Bolivia (AOPEB) and the Italian Association of Biological Agriculture (AIAB) (Trejo, 2008, interview). ANPE is also affiliated to IFOAM and in 2009 ANPE has been elected representative of the Intercontinental Network of Organic Farmers Organizations (INOFO). INOFO is an organization that aims to integrate all organizations of small scale producers worldwide. Currently, it integrates 8 producer associations in Latin America, including ANPE, the Association of ecological producer organizations of Bolivia (AOPEB), the Costa Rican Organic Agriculture Movement (MAOCO) and the organic producers of Guatemala. INOFO has risen due to the common concerns and interests of organic producers such as the trading of their products within Latin America (Trejo, 2010, interview). Finally, ANPE receives financial support from a set of international cooperation agencies. The Dutch Humanist Institute for Cooperation with Developing Countries (HIVOS) supports ANPE since 2001 to promote the sustainable development of small scale producers in Peru, especially to implement the Participatory Guarantee System (PGS) in small scale farms. HIVOS support also the strengthening of ANPE in terms of building capacities to organize small scale producers in Peru and the connection with like-minded NGOs of neighbouring countries (Trejo, 2010, interview). From 2007 to 2011 HIVOS provided funding to cooperate with ANPE to enhance the livelihoods of Andean rural small scale producers in collaboration with IDMA, Arariwa and other NGOs of Ecuador and Bolivia. HIVOS currently (2012) provide funding to ANPE to strengthen the national organization and the associations of ecological small scale producers (HIVOS website, 2012). Also, ANPE gets support from Bread for the World, Aid for Development Gembloux (ADG) and Italian Municipalities (Trejo, 2008, interview). ADG is currently (2012) supporting ANPE to strengthen the food sovereignty in the sub-national region of Ancash. The support of ADG is helping to establish the Bioferia Carhuaz and organize small scale producers in the provinces of Huari, Ayja and Caraz. ANPE links Italian municipalities with Peruvian municipalities to implement the 'healthy municipality' project (Trejo, 2010, interview). It also has connections with the Slow Food movement and Slow Food Foundation, branch Italy, (Appendix 2a) (Alvarado, 2008, interview; Pardo, 2008, interview; Trejo, 2008, interview) and the Italian Association of Biological Agriculture (AIAB). OXFAM-INTERMON has provided funding to ANPE for two years (2010-2012) to implement a project to influence the national policies in order to increase the national investment for small scale producers. This project is implemented in the sub-national regions of Ayacucho, Cusco, Junin and Piura. OXFAM-INTERMON has also financially supported the alliance between CNA, CCP and ANPE (Trejo, 2010, interview) and currently (2012) it is providing funding to ANPE to build capacities in small scale producers (ANPE website, 2012). Currently (2012) ICCO is another agency that collaborates with ANPE as part of a larger collaboration with the Agro-ecological Consortium. In the past the Vredeseilanden (VECO), a Belgian NGO working in Ecuador, has funded a small project of commercialization of organic products in Piura and Lambayeque, two sub-national regions of Peru next to the border with Ecuador, and their connection with Ecuador (Trejo, 2010, interview).

Network relationships: power

The national coordination board of ANPE is the national node of this network. Local nodes are the local producer associations. At global level, key actors are the Humanist Institute for Cooperation with Developing Countries (HIVOS), the Aid for Development Gembloux (AGD) and the International Federation of Organic Agriculture Movements (IFOAM). Instead of regional platforms, sub-national platforms have been identified. The key sub-national nodes are the Association of Organic Producers of Piura (ARPEP) and the Organic Producer Association of Ayacucho (ARPOA).

Before organic producers established ANPE, NGOs affiliated to RAE Peru used to be their unique channel of funding. Now, ANPE is taking over several tasks of RAE Peru, including the funding of organic producers. As a result of this increase in power of organic producers in the networks, conventional NGOs are increasingly becoming accountable not only to their funders but also to their producer beneficiaries. Before 2010 conflicts between NGOs and producers were intense. However, from 2010 onwards conflicts have decreased and cooperation has increased, though in the PIDASSAA platform there are still strong conflicts between conventional NGOs and producer NGOs. Cooperation of conventional NGOs and ANPE is key to influence the national government in order to increase public investment for small scale producers. Another cooperative example is the fact that Arariwa, a NGO based in Cusco, works together with local producers affiliated to ANPE. In this new scenario, professional staff of NGOs helps to develop technical proposals and producer representatives take the final decisions. Similarly, ANPE and IDMA are cooperating to implement the Participatory Guarantee Systems (PGS) for small scale producers (Alvarado, 2008, interview; Trejo, 2008, interview). Therefore, producers need NGOs that facilitate their access to competitive markets. As the roles of NGOs and producers are re-defined, new platforms have emerged such as the Agro-ecological Consortium. The Agro-ecological Consortium is enabling alliances with new organizations, including the Network of Rural Municipalities of Peru (REMURPE). 'It is clear for conventional NGOs that producers themselves have to participate directly in the decisions. Now the roles are defined. Conventional NGOs provide advice to producers, but they do not lead the Peruvian Agro-ecological Movement anymore' (Trejo, 2010, interview).

The ANPE national coordination board and its affiliated organic producers have asymmetrical power position in controlling decision making and agenda setting. This top-down relationship enhances conflicts among organic producers. While producers affiliated to Grupo Ecologica Peru stand up for third party certification, most producers affiliated to ANPE advocate PGS. Sub-national platforms are also the place for disputes between organic producers affiliated and non-affiliated to ANPE. However, the need for cooperation of organic producers is higher, which downplays conflicts at the sub-national platforms. This is shown in the conformation of the Association of Organic Producers of Piura (ARPEP).

Finally, asymmetrical relations between ANPE and HIVOS and ADG are identified. ANPE depends for funding and expertise on those agencies to implement projects. However, the capacity of ANPE and of producers to cooperate create a room for collective initiatives (e.g. Bioferias, Mistura, organic stores, ecotourism) that do not make them fully dependent on funding from international cooperation agencies.

5.2.4 Organic farmers' market in Peru: the Bioferias

In this section, the Bioferias are introduced in order to understand better the networks of organic production in Peru. The Bioferias are the points where key actors of the agro-ecological network, the organic market network and the ecological farming network, particularly NGOs, confluence.

The Bioferias are local organic farmers' markets launched in Peru by agro-ecological NGOs affiliated to RAE Peru in cooperation with organic producers. The Bioferias are marketplaces for producers associations, individual producers and small agro-processors to commercialize their organic products directly to consumers. The Bioferias follow the traditional practice of fairs in the Andean societies to generate a market for organic products. The first Bioferia, located in the city of Huanuco (north of Peru, in the mountains), was established by IDMA in cooperation with ANPE. At the moment there are 22 Bioferias spread over several cities and towns in Peru. NGOs played key roles in the setting up of Bioferias in Peru.

The Bioferia Miraflores

The Bioferia Miraflores is the most important Bioferia of all Bioferias of Peru in terms of the amount of producers, providers and revenues. It was launched in the district of Miraflores, in Lima, the capital of Peru, in March 1999 by Grupo Ecologica Peru. At the beginning, agro-ecological NGOs and producers had to overcome constrains related with the bad reputation of conventional farmers' market at streets in Lima. Although, farmers' markets have been the usual way to 'buy' goods in Peru since ancient times, in modern times, especially in higher income areas of big cities like Lima, they are seen as noisy and chaotic. The Bioferia Miraflores broke those prejudgments and gained the acceptance of middle and high income consumers.

In 2009 the Bioferia Miraflores housed 50 producer representatives, both associations and individuals, organized in 48 stands. Only 15 producers of the 50 producers are affiliated to Grupo

Ecologica Peru. One stand might commercialize products of 10, 20, 100 or 200 producers. For instance, the association Multicielo represents 200 producers of the Lurin valley, located in the south of Lima. The Bioferia Miraflores involves in total about 1000 small scale producers supplying products from 10 sub-national regions of Peru: Piura, Huánuco, Junín, Huancavelica, Ica, Arequipa, Cusco, Tumbes, Ucayali and Lima (Pardo, 2008, interview).

Before the Bioferias, the only option for organic producers to commercialize ecological products at the local market was the 'Biocanasta' (delivery of organic food) but it failed due to the scarce supply of products (Alvarado, 2008, interview; Trejo, 2008, interview). The Bioferia Miraflores has flourished economically due to the higher price of organic products, about 30% more than non-organic products. The location of the Bioferia in a city with high income citizens eager to buy organic food also contributed to its success.

The supply to the Bioferia Miraflores is regulated by Grupo Ecologica Peru. Producers cannot commercialize what they want. Regulation is intended to ensure fair prices, sufficient supply of products and fair competition among producers. It protects the Bioferia Miraflores from over production and concentration of supply only in a few products. Grupo Ecologica Peru in cooperation with the Bioferia Miraflores 'watch dog' group plans the supply of products using a schedule of supply. The watch dog group represents producers of each kind of products, including fresh vegetables, dairy products and grains (Pardo, 2008, interview).

The Bioferia Miraflores has become the model of enterprise development and an opportunity for small scale organic producers to scale up. Several enterprise brands have been nurtured in the Bioferia Miraflores (Table 5.2). For instance, La Cabrita supplies organic food products to the supermarkets WONG and BIBANDA, restaurants and hotels in Lima. The association Biofrute also supplies directly to supermarkets (Wu, 2008, interview). In the last 8 years, producers of the Bioferia Miraflores have gained experience in ecological farming techniques and commercialization strategies. Now affiliated producers are becoming more independent from Grupo Ecologica Peru. For instance, Valle Sano, a producers association of Bioferia Miraflores, aims to become a market intermediary organization like Grupo Ecologica Peru in the future (Alvarado, 2008, interview; Pardo, 2008, interview).

5.2.5 Networks: patterns, challenges and trends

In this section the main patterns, challenges and trends of the agro-ecological network, the organic market network and the ecological farming network are described. The analysis of the main patterns found in the networks focuses on cooperation, power relations and roles of NGOs.

Cooperation

Cooperation of actors in the networks of organic production has the following four characteristics: (1) cooperation implies strong exchange of resources and clear division of tasks; (2) cooperation is flexible; (3) cooperation enhances the expansion of networks; and (4) cooperation is not a direct consequence of the existence of ties.

Regarding the first characteristic, actors exchange resources continuously, either within the networks or with other networks. The exchange of resources is done with particular networks or single actors that fulfill better mutual interests, the so called 'key' partners. In the networks RAE Peru, Grupo Ecologica Peru and ANPE have closer cooperation, among each other, with other NGOs and within platforms (Appendix 2a,b,d). However, cooperation has particularities in the networks of organic production. One particularity is that at local level cooperation rests strongly on non-formal mechanisms while at global level it rests on formal mechanisms (Alvarado, 2008, interview; Trejo, 2008, interview; Wu, 2008, interview). Moreover, cooperation implies task division of actors in the networks. Actors contribute with particular resources in terms of knowledge, information and capital (Alvarado, 2008, interview; Pardo, 2008, interview). For instance, the NGOs Centro IDEAS focuses on trainings and research, IDMA, Huayuna and CEAR focus on ecological farming, RAE Peru focuses on capacity building, local market research and logistical support of producers, RAAA focuses on lobby and advocacy, and Grupo Ecologica Peru and ANPE focus on market access and development respectively (Alvarado, 2008, interview). The networks, the main local actors, platforms and promoters are summarized in Table 5.4.

Regarding the second characteristic, flexibility of cooperation means that actors establish platforms and other operational structures on temporary or continuous bases, depending of the needs in the networks at sub-national, national, regional and global level. NGOs group in task

Table 5.4. Networks promoting organic production for SMEs in Peru (see list of abbreviations for explanation).

Networks/national actors	Platforms				Global actors
	Local	National	Regional	Global	
Agroecological					
RAE Peru and its affiliated NGOs	Bioferia Surco Bioferia Lima Norte Bioferia Cieneguilla	Peruvian Agro-ecological Consortium Peru, Country Free of Transgenics	MAELA GALCI	IFOAM	ICCO
Organic market					
Grupo Ecologica Peru and its affiliated NGOs and small scale enterprises	Bioferia Miraflores Bioferia San Isidro Bioferia Lima Norte	Peru, Country Free of Transgenics	MAELA GALCI	IFOAM	CORDIAD OXFAM-NOVIB
Ecological farming					
ANPE and its affiliated small scale producers	13 Bioferias throughout Peru ARPEP	Peruvian Agro-ecological Consortium Peru, Country Free of Transgenics	MAELA	IFOAM	HIVOS ADG OXFAM-INTERMON INOFO Slow Food

forces, steering committees or working groups for short periods of time or they come together for setting up a long-term new organization. Moreover, NGOs can join or split up to establish a new platform, or join to an already functioning platform.

Regarding the third characteristic, members of the networks of organic production have strong ties, especially among key members. RAE Peru, Grupo Ecologica Peru and ANPE present strong ties, especially RAE Peru. RAE Peru is a more established actor than the other two NGOs in terms of its longer existence, being founded in 1987 while key affiliated NGOs were founded at the end of the 1970s and beginning of the 1980s (Alvarado, 2008; interview; Wu, 2008, interview). Ties are not limited to the networks of organic production. New ties are formed with fair trade, agrarian and business networks, international cooperation agencies, consumers and governmental agencies (Pardo, 2008, interview; Trejo, 2009, interview). Actually, Grupo Ecologica Peru and ANPE are expanding their ties. Especially, Grupo Ecologica Peru is establishing ties with market actors (Pardo, 2008, interview). Furthermore, the struggling against transgenic food is triggering the formation of ties with new actors joining the organic networks, including, exporters, chefs and vegetarians. This expansion of ties sets the condition for furthering cooperation among actors. As a result, cooperation has become more intense in the networks of organic production. This capacity of cooperation of local NGOs in Peru in terms of mobilization of stakeholders and political advocacy is also highlighted by Zelada (2008) in a study of 16 local NGOs counterparts of CORDIAD. An overview of the networks of organic production; including key actors, platforms and connections, is presented in Figure 5.1.

Regarding the fourth characteristic, the presence of ties among actors is not enough to initiate cooperation. This means that although there are connections among actors, cooperation might be scarce or absent. RAE Peru, Grupo Ecologica Peru and ANPE have showed scarce cooperation with other organizations and platforms working on rural SME development, specially before 2008, although they had strong ties (Alvarado, 2008, interview; Pardo, 2008, interview; Trejo, 2008, interview). Furthermore, the scarce cooperation happened also between agro-ecological NGOs and nature conservation NGOs. However, during the last decade the increasing interaction between food production and nature protection is encouraging the cross fertilization of both types of NGOs, especially in the mitigation strategies for climate change.

Power relations

Power relations between actors in the networks of organic production have the following three characteristics: (1) power is expressed in terms of control of resources; (2) key actors are involved in disputes and conflicts for conserving or gaining power positions; and (3) power positions of actors are continuously challenged and change in time.

The control of resources in the networks is operationalised by rules and centralized coordination. Asymmetry in controlling resources is stronger at the local branches of the networks. ANPE has more centralized control of its affiliated organizations than RAE Peru (Alvarado, 2008, interview). For instance, ANPE's coordination board emphasizes direct supervision of their affiliated producers and monitors the work of other actors in the networks. ANPE acts as a watch dog organization, monitoring the way funding should be allocated by other actors, especially by RAE Peru (Alvarado,

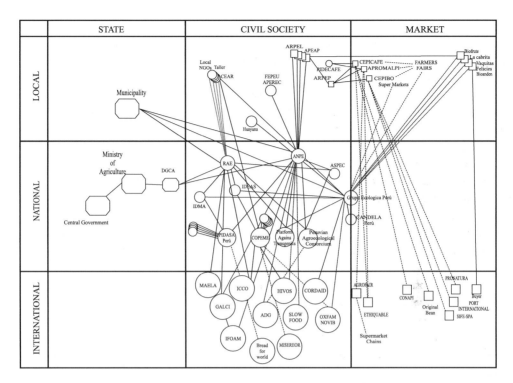

Figure 5.1. Map of the networks of organic production; including key actors, platforms and connections.

2008, interview). At international level, ICCO, HIVOS and OXFAM/NOVIB mainly have the capacity to control the provisioning of funding and expertise to the network.

Regarding the second power characteristic, power struggle has been intensified because of the more central position of producers either forming their own organizations or gaining power in organizations they joined with NGOs. Conflicts and disputes between NGOs and producers have emerged in the networks as a result of the control of resources. Indeed, networks of organic production are far from a harmonious space. Conflicts are usually between NGOs and producers affiliated to Grupo Ecologica Peru, and between RAE Peru and ANPE. The conflicts are mainly between their national coordination boards. Also, conflicts between producers commercializing at the Bioferia Miraflores and Grupo Ecologica Peru (Pardo, 2008, interview). One of the factors found is this study and pointed out by other studies (e.g. Zelada, 2008) is the paternalistic approaches of conventional NGOs to small scale producers, not realizing that small scale producers are the one that run businesses and know better how markets operate. Conflicts between NGOs and producers have to do with the emancipation of producers from NGOs and the need to ensure powerful positions in the network.

As a result of those disputes there are now changes in the position of actors in the networks, the third characteristic. Power relations between producers and NGOs now differ from those in the 1990s (Pardo, 2008, interview; Trejo, 2008, interview). For instance, the setting up of ANPE by producers and the balance in power between producers and NGOs in Grupo Ecologica Peru show that producers are capable to lead their organizations. At the beginning of the Bioferia Miraflores

NGOs and producers representatives agreed that NGOs should lead Grupo Ecologica Peru while producers gain experience. However, NGOs did not leave the board of Grupo Ecologica Peru, losing legitimacy in the eyes of producers. As a consequence, in the year 2004 another turning point happened with the change of the Bioferia Miraflores. Producers of the Bioferia Miraflores raised concern about the fact that Grupo Ecologica Peru is only in the hands of NGOs. Most producers started to see Grupo Ecologica Peru as another organization and did not feel part of Grupo Ecologica Peru anymore. Producers organized themselves to be part of the board of Grupo Ecologica Peru. The same year, for the first time in the history of an NGO in Peru, producers took the lead of Grupo Ecologica Peru (Pardo, 2008, interview). Now, RAE Peru, Grupo Ecologica Peru and ANPE have more horizontal participation at the national level platforms. Another proof of better positions of producers is the emergence of organic food enterprises such as 'La Cabrita', 'Bioandean' and 'Vaquitas Felices'. They are gaining a position in market as the organic market growth. This more central position of producers does not mean the end of NGOs. The sensitivity of local NGOs to the reality of small scale producers makes them necessary in the chain of cooperation from international cooperation agencies towards small scale producers. The capacity of local NGOs and producers to mobilize resources, networking and establish cooperation – despites their disputes – make local actors holders of bargaining power. Particularly, the participation of small scale producers in NGOs and platforms make the networks of organic production socially strong.

Roles of NGOs

NGOs in the networks of organic production perform the following main roles: (1) they support the development of organic markets; (2) they support small scale producers and enterprises to access competitive markets and; (3) they support enterprise development.

Actors in the networks of organic production converge in supporting the Bioferias in Peru. NGOs have tremendously contributed in developing the local organic market, providing funding for the Bioferias, organizing trainings on organic farming techniques and on certification schemes, coordinating with municipalities, and influencing policy makers for institutionalising the Bioferias.

With respect to small scale producer support, the roles of NGOs change. New roles are market facilitation and market mediation of producers and enterprises with market actors. The target is supermarkets, restaurants, hotels and caterings at the national level, and big stores and supermarket chains at the international market. As conventional NGOs have limitations to answer the new needs of small scale producers in the market, this gap has contributed to the rise of a new type of NGOs. Examples of such new types of NGOs are Grupo Ecologica Peru and ANPE. It is important to point out that a more central position of small scale producers and enterprises in the market does not mean a diminishing role of NGOs. NGOs are getting further expertise in combining the development of and access to organic markets. For instance, the NGO El Taller is setting up a new Bioferia in Arequipa, a big city in the south of Peru, and supporting small scale producers of organic medicinal plants to access international markets (Alvarado, 2008, interview; Luque, 2007, interview).

Thirdly, NGOs are adjusting their traditional roles towards the new context of the liberal market. NGOs are providing capacity building to small scale enterprises in business skills, market information for producers, consumer demands, business management and certification schemes.

Challenges

Networks of organic production are facing three main challenges: (1) organizational sustainability; (2) trust between actors; and (3) meeting organic market demands.

The first challenge has to do with the concern of NGOs for their sustainability. NGOs understand sustainability challenges at three levels: individual, the platforms and the entire network. One of the main concerns highlighted by NGOs is funding. NGOs financially depend strongly on international cooperation agencies to operate projects with producers. Nevertheless, dependency of aid funding is not the same for all NGOs. Centro IDEAS and IDMA have higher dependency from international cooperation agencies than ANPE and Grupo Ecologica Peru. As the last two NGOs are involved in commercialization of organic products, they depend also on the market. The availability of funding for conventional NGOs is getting scarce. In the case of the new type of NGOs, including producer NGOs and market NGOs, there is more funding available for their projects. However, they are not enough prepared to take that opportunity. Similarly, the Bioferias are facing challenges regarding financial sustainability. Before 2001, the Bioferia Miraflores was a subsidized initiative by RAE Peru. Now under the market demand approach, producers have to take into account the commercialization costs at the Bioferia, for instance, paying administrative costs, taxes and logistic costs. Producers claim to be more independent from NGOs but at the same time want to remain subsidized by NGOs. The Bioferias also show limitations for business growth of producers. The quantity of supply is low and there is no room for the upgrading enterprises. For instance, the Bioferia Miraflores is overcoming its capacities in terms of size, space and consumers affluence. According to Pardo (2008, interview), at least 200 Bioferias are needed in Peru to put a respectful quantity of organic production on the market.

The second challenge consist of overcoming the mistrust especially between NGOs and the national government, and between producer NGOs, market NGOs and conventional NGOs. Key actors of the networks of organic production and of the national government lack strong ties. Although some ties have been identified with SENASA and the Ministry of Agriculture, no lasting cooperation has been identified.

The third challenge has to do with the demand for knowledge on market-oriented organic farming and on strategies of commercialization. Organizing production according to market demands means understanding how the market works and how to commercialize products. The possibility that small scale organic producers supply products to local and international markets is an opportunity, but most small scale producers are not prepared to be suppliers in terms of quantity, quality and continuity. Most small scale organic producers have limitations in terms of business competences. Similarly, agro-ecological NGOs need to develop capabilities for market facilitation and market intermediation. Finally, three basic requirements have been raised in the networks to meet market demands: increase the number of organic producers, the scale up of production and the adoption of sustainable production technologies. What has been found is that the promotion of sustainable technologies is not a priority, neither for agro-ecological NGOs nor for the national government. Furthermore, national agrarian policies do not establish enabling conditions and do not ensure enough business capacity for the growth of the organic market in the country. The local organic market can be enlarged by using PGS but there is lack of governmental support to do so. The entering of large agro-industries into the local market will also impact the production of small scale organic producers. By having capital and getting support from government, large agro-industries can overtake small scale producers in the organic market.

Trends

The networks of organic production show the following main tendencies: (1) change in values, (2) further density of connections and (3) further formalization of local actors.

Values are changing in the networks, from solidarity to business, due to a more central position of the neoliberal market. Actors and platforms are being organized in function of market demands. This process is clearly observed in the Bioferia Miraflores. At the beginning the Bioferia Miraflores did not focus on profit making. Before the 1990s, motivation of NGOs and producers was driven by solidarity values. For instance, there was close cooperation between Grupo Ecologica Peru and the Centro IDEAS. The consequences of the change in values were the formalization of Grupo Ecologica Peru and the Bioferia Miraflores in 2001. NGOs and organic producers are centralizing the Bioferias into a national platform. In this way, it is expected that producers supply products to several Bioferias. Additionally, actors in the networks are increasingly using management, monitoring and evaluation tools to measure the performance of their projects. Professionalization is becoming a common practice, for producers by bringing their products in a competitive market and for local NGOs by becoming attached to the chain of international cooperation agencies. Networks of organic production will likely become more structured at regional, national and sub-national level, with stronger ties and highly structured platforms.

5.2.6 Conclusions

a. The agro-ecological network, the organic market network and the ecological farming network are the main networks involved in the promotion of organic production by small scale producers and small scale enterprises in Peru. The three networks are mainly constituted by agro-ecological NGOs, small scale organic producers and international cooperation agencies. Each network engages a particular set of actors from local to global level. Agro-ecological NGOs are the main actors in the networks. Key agro-ecological NGOs in Peru are RAE Peru, Grupo Ecologica Peru and ANPE. They are not single organizations but a 'cluster' of local organizations, including their affiliated organizations operating throughout Peru and their national coordination boards located in Lima, the capital of Peru. RAE Peru represents most agro-ecological NGOs in Peru. Grupo Ecologica Peru represents agro-ecological NGOs and small scale organic producers. ANPE represents only organic producers, not NGOs. Different from Grupo Ecologica Peru and RAE Peru, ANPE is organized in sub-national platforms of producers, being the Association of Organic Producers of Piura (ARPEP) and the Organic Producer Association of Ayacucho (ARPOA) the emblematic examples.

b. Agro-ecological NGOs are organized in platforms established at local, national, regional and global level. The platforms are hubs for the coordination of actors. The first layer is the local platform of the organic farmers' market called Bioferias. Bioferias are widely spread throughout the country. The second layer is the national platforms, the best known being the Platform 'Peru, Country Free of Transgenics' and the Agro-ecological Consortium. The third layer is the regional platforms. Examples are MAELA and GALCI. The fourth layer is the platform of global reach such as IFOAM. Exchange of resources and cooperation between actors in terms of funding and expertise are strong in the platforms. Platforms can be established for long-term or short-term. For instance, while the Peruvian Agro-ecological Consortium has been launched, the National Commission of Organic Products (CONAPO) has been dissolved. Platforms are the nodes where agro-ecological NGOs build ties with non-agro-ecological NGOs, governmental actors, agrarian associations and business actors. Actors expand or reduce their ties according to their needs, although ties are very strong in the networks. This means that cooperation is intense in the networks but flexible. As the rise of the neoliberal market has brought new needs for actors in the networks of organic production, the networks are expanding their boundaries, creating new connections with new actors.

c. NGOs and producers are involved in disputes and conflicts for power positions in the networks. Different motivations, interests and limited access of funding mainly condition the emergence of conflicts. Most conflicts are related to the power position of producers in the platforms and the control of producers by their national coordination boards. This puts producers in a dilemma between the need for cooperation with NGOs and the need for gaining power. The current expansion of the local organic market, especially the Bioferia Miraflores, the supply to supermarkets and the mistrust between producers and conventional NGOs are also enhancing conflicts. However, the need of organic enterprises for expanding their business and reaching new markets, and the will of NGOs to support small enterprises make cooperation stronger than conflicts. Power is continuously challenged in the networks, originating changes in the power position of actors. While the power position of RAE Peru was central to develop organic production in Peru during the last decade, currently Grupo Ecologica Peru and ANPE perform the more central positions. In addition, the networks are changing in their values from solidarity to business and local actors are moving to further formalization. Besides mistrust, the networks of organic production are facing other challenges, including the sustainability of key NGOs and platforms, and the scarce business capabilities of organic enterprises.

d. Three types of agro-ecological NGOs have been identified; conventional NGOs, market NGOs and producer NGOs. Each one performs particular roles in promoting organic production. Conventional NGOs are NGOs with a long history working on supporting small scale producers and local organic market development. The role of conventional NGOs has been central in making possible the achievements of organic production in Peru. RAE Peru is an example of a conventional NGO. Market NGOs are NGOs performing intermediary roles at local markets. Market NGOs have built ties with market actors. Market NGOs are 'dual' organizations that are not only supporting small scale producers but also are intermediating them to access local and international markets. Market NGOs have emerged to answer the needs of producers for specialized knowledge on commercialization, business skills, marketing strategies, demand creation and product development. Grupo Ecologica Peru is an example of a market NGO. Producer NGOs are NGOs that are constituted by small scale organic producers. Producer NGOs focus on improving basic techniques of production and commercialization, and influence governmental agencies to develop public policies in favor of organic production. Producer NGOs rose as a result of emancipation of producers from conventional NGOs. ANPE is an example of a producer NGO. Conventional NGOs and producer NGOs have differences and similarities. Both need funding to keep the organization operating and to finance projects for their affiliated organizations. The difference is that producer NGOs not only work like NGOs but also like unions, being also 'dual' organizations. The fact that producers bring their products at the market makes their organizations less dependent of funding from conventional NGOs.

5.3 Discourse analysis of organic production

5.3.1 Introduction

Agents are basic elements of discourses and central actors of networks. Agents subscribing the discourses of organic production are part of the agro-ecological network, the organic market network or the ecological farming network. It should be kept in mind that organic discourses do not match necessarily one-to-one with the networks of organic production.

Based on the Dryzek's (1997) definition of discourse, organic production can be seen as a diverse set of assumptions, judgments and contentions. Discourses of organic production are shared ways to apprehending organic food production and commercialization encompassing plausible storylines or accounts about market and sustainability of NGOs, small scale producers, consumers and the national government. The discourse analysis of organic production focuses on the views about market and sustainability, and the role of key agents involved in organic production and commercialization, among which local and international NGOs, organic producers and the national government.

The discourses of organic production are analyzed following the methodology of Dryzek (1997), as described in depth in Chapter 3. Four elements are analyzed to construct storylines: (1) basic entities acknowledged; (2) assumptions about natural relationships; (3) agents and their motives; and (4) key metaphors and other rhetorical devices.

Discourses of organic production

The organic production discourses identified in this research can be clustered in two groups. The first group includes the discourses centered on the market and the second group the discourses centered on self-sufficiency.

The first group of discourses centered on the market includes two discourses stressing either adaptation or access of small scale producers to markets. 'the market adaptation discourse' embodies the idea that actors in organic production, especially 'traditional' NGOs and small scale producers, are forced to gain new capacities to adapt successfully the overwhelming market economy. NGOs take the leading role since small scale producers do not have the competences to take the challenge of adaptation by themselves. In this light, small scale producers need support from agents to adapt to the neoliberal market. Support is understood in terms of transferring resources, especially capital and knowledge, capacity building and awareness rising from agents to small scale producers. The market adaptation discourse highlights the setting of models of small scale commercialization, especially Bioferias and Biotiendas as a more suitable strategy to nurture organic enterprises. The 'market access discourse' addresses the idea that small scale producers are eager to move to competitive markets. Despite the fact that small scale producers need support of agents, they play a leading role in this endeavor. In this light, rather than just broad capacity building strategies, small scale producers need support in specialized issues such as organic farming technologies, financial management and market information to organize supply to competitive local and international markets. 'Market' NGOs are the specialized market-oriented agents subscribing to this discourse. To do so, small scale producers are active agents building up their own support organizations and approaching potential new partners. In this discourse the channel of commercialization are profitable markets such as Bioferias in high income cities and supply to local and international large supermarkets and organic food shops.

The discourses in the second group endorse food sovereignty and underline the notion that food production of small scale producers should mainly be oriented to self-consumption. And only the surplus should be commercialized at local markets. The two discourses within this group are different in the sense that while the 'market democratization discourse' stands for agro-ecology, the 'peasant movement discourse' stands for traditional and sustainable agriculture. The market democratization discourse highlights the widening of organic markets for low and middle income consumers. It claims that producers should sell directly their products to consumers or buyers, without the need of market intermediaries. Rather than adapt or access to the liberal market, small scale producers intend to build up a fair relationship with it. In the long-term small scale producers also seek to reach competitive markets as a result of a wider social change. The main agents in this discourse are 'producer' NGOs. In the case of the peasant movement discourse, it emphasizes the social and cultural values and traditional practices of food production. The peasant movement discourse rests on resisting and opposing the liberal market economy and struggling against the national government. The peasant movement discourse contests the liberal market economy and encompasses income generation for small scale producers in small local farmers' markets in the short-term. However, the issue how to scale up traditional production at the long-term is not clear in this discourse. Regarding the strategies of commercialization, the market democratization discourse claims that the production surplus at organic farmers' market need an organic guaranty scheme, while the peasant movement discourse stresses traditional farmers' markets without any assurance scheme for their products. In Table 5.5 an overview of the three discourses addressing organic production are presented.

In the next sections, the discourses of market adaptation, of market access and of market democratization are developed in detail.

5.3.2 Discourse of market adaptation

Let's adapt to liberal market!

The discourse of market adaptation claims that actors, especially NGOs and small scale producers linked to the organic production, are forced to obtain new capacities and be well equipped to adapt to the liberal market economy. Otherwise, they will be excluded. Adaptation means that despite NGOs and small scale producers adjust to the new market reality. By promoting models of small scale commercialization such as organic farmers' market and organic shops, NGOs nurture organic producers and enterprises. An overview of the analysis of the discourse of market adaptation is at Table 5.6.

Table 5.5. Overview of discourses on organic production.

Discourse	Adaptation to liberal market	Access to liberal market	Market democratization
Storylines	NGOs and small scale producers are forced to get new capacities and to adapt to liberal market. Small scale producers do not have the competences to take the adaptation challenge by themselves. Agents nurture organic producers and enterprises.	Small scale producers are eager to move to competitive markets. Support is needed from specialized agents in managerial and technological issues to organize supply to competitive local and international organic markets.	Prioritize organic market for low and medium income consumers. Rather than adapt or access to liberal market, small scale producers intend to build up a fair relationship with it.

Type of NGOs	Agro-ecological NGOs		
	Traditional	**Market**	**Producer**
Subscriber	Centro IDEAS, IDMA, CEAR	Grupo Ecologica Peru, PIDECAFE, Candela Peru, El Taller	ANPE, CEPICAFE, APROMALPI, CEPIBO

Basic entities acknowledged

The liberal market is acknowledged and its worldwide spreading is pushing actors of organic production for change. Actors, especially NGOs and small scale producers, are forced to gain new capacities to fulfill new requirements of the liberal market. In that way, NGOs subscribing this discourse realize the importance of markets. Silvia Wu (2010, interview), a member of RAE Peru stated 'improving the production of small scale producers does not have the expected results of satisfying the family basic needs such as a income, education, health services and housing, if they do not know where to commercialize their products'. There is no option for small scale producers to be away from liberal markets in order to satisfy such needs. Discourse subscribers claim that small scale producers necessarily have to bring their products to competitive markets to get a decent income. Therefore, small scale producers have to be well-equipped with adequate instruments to survive the liberal market forces. However, the adaptation of actors to liberal markets does not imply abandoning their former roles fully. NGOs do not necessarily abandon their traditional core views, neither do small scale producers abandon producing for self-consumption. Traditional core

Table 5.6. Discourse analysis of market adaptation.

Elements	Market adaptation
Basic entities acknowledged	Liberal market
	Organic market and consumption
	Commercialization channels: Bioferias and Biotiendas
	Sustainability
	Organic and non small scale organic producers
	Local networks of NGOs and their international partners
	Organic consumers
	National government
Assumptions about natural in relationships	Solidarity
	Hierarchy
	Rivalry
	Mistrust towards the national government
	Nature guaranties human society survival
Agents and their motives	NGOs: from paternalism to market facilitators
	Producers: beneficiaries
	Consumers: rational, individualistic
	State: bureaucratic, opportunistic
Key metaphors and other rhetorical devices	Conventional NGOs: equippers of producers
	Small scale producers: weak, immature people and competitors
	Liberal market: powerful force
	Business is dirty
	Market: battle field between organic and modern agricultural promoters
	Sustainability: holistic human body system with specialized organs
	National government: unpredictable, opportunistic
	Rhetorical device: capacity building, participation and beneficiaries

views of NGOs have to do with standing as aid organization and seeing business and commercial roles as non-trustful and complex endeavours. Similarly, production for self-consumption means to ensure high quality food and safeguarding nutrition of producer families in the rural areas. The pressure of the liberal market economy is also forcing conventional NGOs to diversify their funding sources and pushing small scale producers to adapt to the liberal market. This is the case for the Centro IDEAS that is providing support to producers on new issues, including the elaboration of business plans, marketing, product development and commercialization strategies, however without abandoning its traditional core viewpoints (Alvarado, 2008, interview).

Organic markets and consumption have also been acknowledged in the discourse but to a limited extent. Attention is given to creating organic local markets, limited to organic farmers' market (the so called 'Bioferias') and organic shops (the so called 'Biotiendas') in high income areas of big cities. International markets are considered not reachable and low income towns are considered not profitable. The discourse considers consumers key actor for the development of organic markets. The Bioferias are also considered a truly business school for producers to learn about market and consumer preferences. Much of the current supply of organic products at local markets is the result of the work of NGOs affiliated to RAE Peru and RAAA during the last two decades (Alvarado, 2008, interview; Wu, 2008, interview).

Sustainability is understood as consisting of multidimensional pathways. In individual terms it is seen as a holistic 'green' lifestyle. In organizational terms, sustainability means diversification of funding sources to promote organic production. Organic production is feasible as far as the sustainability of natural resources is guaranteed. Emphasis is given to certain approaches of sustainability. For instance, while some agro-ecological NGOs emphasize socio-economic concerns, others emphasize both socio-economic and ecological concerns. For example, the Centro IDEAS, CEAR and IDMA highlight the former concern, and AEDES and RAAA highlight the latter concern. In any case all NGOs subscribing to this discourse consider sustainable development as the guiding concept for small scale producers and rural development. Sustainability of producers is seen linked to the protection of nature.

Other basic entities recognized in the discourse are international cooperation agencies, the Peruvian government and organic consumers. They are described in detail in the section agents and their motives.

Assumptions about natural relationships

Solidarity is a characteristic value that mediates the relationship among actors subscribing to the market adaptation discourse. In the market adaptation discourse, solidarity has a wide range of explicit social connotations such as 'gratitude', 'help' or 'pity' towards 'unprivileged' small scale producers. Considering small scale producers as 'unprivileged' people is a common view of the discourse proponents.

Hierarchy – especially of NGOs – is based on social status and expertise. Indeed, development plans are designed by NGOs to support 'poor' small scale producers and local organic market development. Small scale producers are seen as scarcely resourced and are considered subjects of development interventions.

Rivalry and power struggle are usual between conventional NGOs and producers. They compete for power and funding in the networks. Mistrust and the claim to be a legitimate representative of small scale producers enhance rivalry. Much of the conflicts among actors in the networks of organic production are linked to the formation of market NGOs and producer NGOs, next to the well-established conventional NGOs. The relation of NGOs is involved by strong mistrust and power struggles, especially between professional driven NGOs and producer driven NGOs. In words of Silvia Wu and Fernando Alvarado (2010, interviews) 'the conflict is because of producers are against professional views'. An example of this statement is the conflict and power struggle between Centro IDEAS and Grupo Ecologica Peru. During the period 2007-2008, the Centro IDEAS acted as watch dog of the Grupo Ecologica Peru's former coordination board supervising the expenditures and encouraging the Bioferia Miraflores producers to monitor Grupo Ecologica Peru. In addition, RAE Peru has also disagreed with the way producer board members of Grupo

Ecologica Peru manage the Bioferia Miraflores. In 2009 RAE Peru has set up a new Bioferia in the district of San Borja with a more decentralized administration (Alvarado, 2008, interview; Pardo, 2008, interview; Wu, 2008, interview). Questioning the role of Grupo Ecologica Peru, RAE Peru alleges that the role of NGOs is to facilitate and give advice, but the administration of the Bioferias corresponds to producers themselves. In the words of RAE Peru's director 'Grupo Ecologica Peru is self-sustaining financially the supply to supermarkets from the Bioferia Miraflores ravenous. Grupo Ecologica Peru has opened a new Bioferia in the district of San Borja, abandoning the Bioferia San Isidro'. This clashing of perspectives in Grupo Ecologica Peru almost caused the split up of their affiliated NGOs in 2010 (Alvarado, 2010, interview).

Mistrust towards the national government is the attitude of most actors, especially towards agricultural governmental agencies. NGOs claim that governmental agencies, including MINAG, SENASA and INIA, ignore their efforts in supporting organic production in Peru, especially in the conversion of small scale producers to organic farming. NGOs claim to be the pioneers and leading promoters of organic production in Peru, without any governmental support (Alvarado, 2008, interview; Wu, 2008, interview). Governmental agencies are considered opportunists and incapable to work in partnership and in participatory ways. As Wu (2008, interview) stated 'now that NGOs have worked for years in developing organic production in the country, governmental agencies show up and intend to be the leaders'.

Regarding the relationship with nature, nature is seen as the sustenance base for human survival. As Wu (2008, interview) stated 'nature will give us high quality food – it means ecological food – forever, as far as the environment is not harmed'. The ultimate aim of organic production for market adaptation proponents is not nature protection but socio-economic development of human society, especially small scale producers. This puts the market adaptation discourse in line with anthropocentric views of nature.

Agents and their motives

Agency is granted to both collective and individual actors. NGOs are primary agents in the networks. RAE Peru, an 'umbrella' NGO clustering individual NGOs including Centro IDEAS, IDMA and CEAR, is a key agent of the discourse. In developing market competences for their beneficiaries, NGOs subscribing this discourse are evolving in their roles from promoting farming for self-consumption to promoting farming for the market. The role of organic local market development is central for NGOs subscribing the discourse. For example, since its foundation in the 1980s, RAE Peru provides workshops on organic agriculture, and networking and publications on organic agriculture, for small scale producers to shift from conventional to organic agriculture. NGOs are also taking up roles of networking and facilitating contacts and information for commercial purposes. NGOs subscribing the discourse handle technological and managerial knowledge of organic production and market in broader terms rather than specialized ones. This means that NGOs are a sort of 'chaperons' of small scale producers that are 'acclimatizing' to the liberal market. From the 1990s onwards Centro IDEAS, IDMA, CEAR and other conventional NGOs establish models of organic commercialization strategies at local level, such as the aforementioned Bioferias, organize producers in associations nurturing small organic enterprises and give them logistic support and capacity building (Alvarado, 2008, interview). This widening of roles of conventional NGOs does not necessarily mean that they are fully abandoning their initial views of self-consumption. NGOs subscribing to this discourse use strong lobby and advocacy strategies to influence and confront the national government. For instance, conventional NGOs have played a central role in influencing the national government to establish the National Commission of Organic Products (CONAPO) in 2001 and the approval of the law N°29196 'Promotion of Organic Agriculture' in 2008.

Given their solidarity values, NGOs consider themselves protectors of 'disadvantaged' producers. NGOs consider helping small scale producers as part of their responsibilities, and these NGOs argue that they have the knowledge and the power to do so. According to the discourse proponents, the setting up of Bioferias has been motivated by solidarity values towards poor producers. Conventional NGOs have also supported the initial stages of the development of small organic food enterprises (Table 5.2).

Poor small scale producers have limited agency. While single and disorganized small holders and their families are seen as passive beneficiaries of developmental interventions, self-organized small scale producers, including associations, cooperatives and NGOs, are seen as active 'friendly' or 'conflictive' agents. Friendly producers are identified as partner producers aware of their limitations and capable to understand 'the idiom' of NGOs. 'Conflictive' producers are seen as the cause of conflicts between producers and conventional NGOs. Producers are considered incapable to handle power and money in a responsibly way at the market (Alvarado, 2008, interview; Wu, 2008, interview). The producer leaders of ANPE and Grupo Ecologica Peru are labeled by the discourse proponents as conflictive people.

Larger producer associations and farmer unions are considered agents but with different motivations and worldviews than agro-ecological NGOs. Within a large number of non-small scale organic producers affiliated, larger producer associations are oriented to international markets. In contrast, NGOs efforts are oriented at developing local organic markets. Larger producers are considered by proponents of this discourse as laggards in sustainable production. This is the case of JNC, the largest union of Peruvian coffee exporters, that does not want to invest in building a local market for coffee in Peru (Alvarado, 2008, interview). Larger producer associations usually get support from banks, investors and buyers to develop their business, and not from agro-ecological NGOs. An example is COCLA, a Peruvian large coffee cooperative that supplies coffee to CRAFT general foods. CRAFT is one of the larger global buyers of coffee and represents 20.8% of the global market. COCLA has received support from CRAFT to start producing organic coffee as CRAFT organized COCLA's producers to supply organic coffee for international organic market. The drive for COCLA producers to become organic has been the demand of CRAFT. Agro-ecological NGOs did not play any role in the conversion of production of COCLA producers towards organic (Alvarado, 2008, interview). Larger producer associations are seen as agents with enough resources, organizational infrastructure and networks of support.

In the discourse consumers are seen as rational and individualistic agents. In consuming organic food, consumers look at their own needs, not that of others. Consumer motivations are usually related to health and nutrition. In this sense, consumers do not distinguish certified from non-certified organic food as both are labeled as 'healthy food'. Social and environmental considerations of organic food consumption play only a marginal role. Consuming organic food is considered a matter of individual or family life-style, not a collective social concern.

The national government can be seen as bureaucratic institution. The national government gives little attention to agro-ecological NGOs, even less to producers, because organic agriculture is not seen as a governmental priority. Small scale organic producers are producing and supplying organic food without much support of the national government (Wu, 2008, interview). As the international organic market has become a major business opportunity, the national government intends to play a more central role in organic production. In the eyes of the discourse proponents the national government is opportunistic, and they consider that NGOs alone have contributed to the maturation of organic agriculture in Peru. An example of the low cooperation between NGOs and the national governmental agencies was the 'I Convention of Organic Agriculture: Peru an organic country' in 2008. The Convention was organized by the National Service of Agrarian Health (SENASA) in alliance with the Peruvian Exporters Association (ADEX), but agro-ecological NGOs were not invited. Another example of the scarce effort of the national government to promote organic production is the slow implementation of the National Council of Organic Products (new CONAPO). The new CONAPO might represent a more institutional way of communication between agro-ecological NGOs and the national government, beyond the personal good-will of governmental officers. However, the Ministry of Agriculture does not make it operational up to now (2012). Additionally, the government is referred as 'blind' in the discourse adherents since it does not realize the advantage of organic production for local economy development.

Key metaphors and their rhetorical devices

The market adaptation discourse is rich in metaphors. These include the notion of 'equipment' to refer to all the competences that small scale producers need when they move to the liberal market domains. Equippers are NGOs and other agencies that provide competence tools to face market challenges.

Another set of favourite metaphors referred to the market adaptation discourse to label small scale producers are: 'weak', 'immature people' and 'competitors'. The first label refers to the 'powerless' condition of small scale producers. The second label refers to their lack of experience in dealing with competitive markets. The third label refers to small scale producers organized in their own organizations organized in associations, in cooperatives and/or in NGOs. They are not considered beneficiaries anymore, but 'competitors' of funding, power and producer beneficiaries.

Other metaphors intend to capture the nature of the market. The liberal market economy is characterized as a powerful force moving like a 'tsunami' or 'big wave' that might be harmful for producers (Wu, 2008, interview). Related to these views of the market is also the idea of 'battle field' where, for instance, the promoters of organic agriculture and the promoters of 'conventional' agriculture are the 'fighters'. In this struggle, consumers are seen as allies of conventional NGOs. Neither producers nor academia are as important as consumers, since the latter are the ones who decide whether or not to consume organic food.

Business rational is considered as something 'dirty' that can contaminate producers once they enter the market. In the words of Wu (2008, interview) 'the liberal market contaminates producers, just because it put together money and power. As there is currently no other way to ensure their income generation, producers have to be well-equipped to survive in the market'.

The notion of sustainability interlinks holistically everything, like in a human body. In human bodies different organs each have particular functions; in a similar way sustainability has to be approached from different corners. Some organizations focus on the natural world, others on the social world. With certain emphasis to one or another, NGOs and producers combine the care for nature and agriculture complementarily.

In referring to the national government, the metaphors of 'unpredictable' and 'opportunistic' have occasionally been raised. Actors do not plan ahead with the national government since it is doubtful whether agreements will be complied or implemented properly. However, the usual unexpected political changes at governmental agencies turn down achievements. For instance, a few years ago well qualified professionals in organic production were fired in SENASA after changes at the national government (Alvarado, 2008, interview; Wu, 2008, interview).

A key rhetorical device in this discourse of organic production is the term 'capacity building'. This term cannot be missed in any developmental project towards small scale producers. Capacity building assumes hierarchical relations between NGOs and producers. NGOs build capacities for 'disadvantaged' small scale producers to perform successfully at the market. Capacity building can usually be either in production or market issues. At the production side, capacity building is related among others to handling farming techniques. At the market side, it is related among others to developing business skills and competences. Despite the use of participatory approaches, capacity building keeps being a top-down approach promoted by the market adaptation discourse proponents. Another rhetorical device linked to the developmental role of NGOs is the term 'beneficiaries'. Beneficiaries can be local NGOs, small scale producers or others that are subject of support.

5.3.3 Discourse of market access

Let's access competitive markets!

Small scale producers are eager to move to competitive markets and they play a central role in this endeavour by building up their own organizations and partnerships. Small scale producers are determined to conquer profitable markets. Rather than just capacity building, small scale producers and enterprises need support of specialized agents in managerial and technological knowledge to organize supply to competitive local and international organic markets. An overview of the analysis of the discourse of market access is provided in Table 5.7.

Basic entities acknowledged

Globalization and the liberal market economy are acknowledged as the unique playground for goods and services exchange. This means that small scale producers have to reach highly profitable markets. Proponents argue that the focus on only farmers' markets, including the Bioferias and small organic shops, reduces producers to marginal markets. In the words of Pardo (2008, interview) 'small scale producers are eager to catch up the new trend of the global liberal market'. Small scale producers build up their own support organizations and partners to access markets. The market access discourse encompasses the entering of small scale organic producers to

Table 5.7. Discourse analysis of market access.

Elements	Market access
Basic entities acknowledged	Liberal market economy
	Profitable organic markets
	Technological and managerial competences
	Third party organic certification
	Environmental concern, agro-ecosystems and natural resources are recognized in limited extension
	Sustainability is recognized in economic terms
	International cooperation agencies, national government and organic consumers
Assumptions about natural in relationships	Hierarchy in human society: Technocratic organizations at the top position
	Group based-individualism
	Relationship human society with nature is hierarchical, individualist and anthropocentric
Agents and their motives	Capable and brave organic producers
	Market facilitators, market intermediaries, technocratic organizations
	Sophisticated and individualistic consumers
	Incapable government
Key metaphors and other rhetorical devices	Decoding market information for producers
	Market NGOs: decoders of complex market information for producers
	Business rationality is benign
	Business and market actors ally in combating poverty
	Market: supernatural force
	National government: incapable entity
	Rhetorical device: real life, real world

profitable local and international markets in less harmful conditions by providing them specialized managerial and technological knowledge. According to market access proponents rather than opposing the globalization of markets, the fair access to market should be guaranteed. In this sense, organic markets under the liberal market economy are recognized and they are considered an opportunity to improve the socio-economic conditions of small scale producers. Markets are the driving force of organic production and being excluded from them means for producers to be marginalized.

The concern of job generation is central in the market access discourse. In entering profitable markets at local and international levels, producers are able to generate income and to improve their economic and social conditions. As low income people are not able to pay the higher price that organic enterprises demand for their products, small scale producers should orient their organic products to profitable markets. Subscribers to this discourse argue that once profitable markets are taken, they will move to lower income consumers. Actually, the organic farmers' market in high income areas of Lima and the supply to large buyers and retailers for local and international markets are allowing economic growth of a diverse set of organic enterprises. This growth of organic enterprises is possible as long as a secured profitable market is in place.

Globalizing liberal markets push actors to move for change and innovation. Under the market access discourse, new actors and new types of organizations emerge. Therefore, new specialized skills are required. These skills are defined in terms of tools, technology, proper organizational management and specialized assistance. Specialized skills are required for production and commercialization processes in order to answer successfully to the demands of competitive organic markets. Without considering these specialized competences, small scale producers are not able to access liberal markets in a good position. As Pardo stated (2008, interview) 'organic producers will not be able to perform successfully in competitive markets, if, for instance, they do not increase quantitatively their production and do not work in large scale as agro-industries do'.

The market access discourse acknowledges third party organic certification as a certification scheme to commercialize organic products at local and international markets. This type of certification is provided by certification bodies, private companies specialized in implementing and monitoring the process of organic production.

Environmental concerns, agro-ecosystems and natural resources are not acknowledged fully. They are subordinated to the requirements of organic certification standards and the market. Nature and native crops are counted in the discourse as far as their protection generates income. Markets decide what crops have to be produced and commercialized. The market access discourse concentrates on people rather than on nature. As commercialization become the central issue in the discourse, concerns about price and supply become central in decision making of actors. Proponents take as granted the fact that healthy nature makes organic agriculture feasible. Overall protecting nature, for instance, against extractive industrial and urban pollution, is seen as beyond organic actors' responsibilities. The generation of income is the main motivation to farm organic food, not the protection of local natural resources. Environmental protection is feasible as far as it is connected to social concerns (Pardo, 2008, interview).

Sustainability is acknowledged within economic boundaries. It is understood as improvement of economic conditions of small scale producers and the structuring of feasible organizational strategies that intermediate their access to profitable markets. Sustainability is also understood in terms of rural-urban ties as it happens in the Bioferia Miraflores. Sustainability is also referred to as the organizational capability to handle conflicts among producers.

Assumptions about natural relationships

The relationship assumed by the market access discourse to be most natural is hierarchy. At the top position are technocratic organizations, including market NGOs, companies and producer associations, with expertise and specialized knowledge that assist small scale producers to profitable markets. Charity-like organizations and organic producers producing for self-consumption are seen at the bottom. Technocratic and market driven organizations in organic production emerge as an answer to market needs of small scale producers. These new organizations have challenged the position of conventional NGOs. Conventional NGOs are seen as incapable to fulfil the liberal market needs. In the market access discourse, hierarchy is conditioned by opposing two options dualities: marginal versus central, volunteer versus profitable, and excluded versus included. Being central, profitable and included mean to be part of the liberal market economy and to enjoy its benefits.

Relationships among actors are founded in individualism. Cooperation is based on individual or group gainings rather than collective welfare. NGOs, producers, consumers and the national government see each other as self-interested groups. Producers subscribing to this discourse care much about individual economic achievements rather than collective benefits. Although these groups are continuously interacting, communication does not get to the point of real dialogue. Each actor has its own agenda. 'Common' interests are those of the group, not of the whole society. Then, clashes of interests are unavoidable. Actors are involved in conflicts, especially local NGOs and producers. Self-interests and mistrust are at the root of conflicts. In this light, Peruvian society is seen as a collection of distrustful groups fighting for their (own) group interests.

The relationship of human society with nature is hierarchical, individualistic and anthropocentric. Markets are the driving forces for organic production. At markets, the economic side of organic production becomes more relevant than the ecological side. Ecological considerations are understood in terms of packaging and waste disposal. However, discourse proponents claim ecological considerations have wider implications. For instance, commercialization of organic products has been questioned as not always environmental friendly due to long distance transportation. This is taken by market access components as challenges to be overcome by technology. Ecosystem services are taken for granted. The ultimate aim of organic production is the socio-economic development of people. People undertake environmental protection as responsible consumers or producers on an individual bases, not as an institutionalized interest.

Agents and their motives

Key agents in the market access discourse are companies, cooperatives, producer associations, market NGOs and technocratic NGOs. Key agents can perform as market facilitating organizations, market intermediary organizations and buyers. In the market access discourse these agents have specialized knowledge and expertise in farming techniques, food storage, processing, market and price information, value chains and management.

Market facilitating organizations are 'brokers' connecting organic producers with economic actors. Market facilitating organizations make feasible the supply of organic products to buyers. Market facilitating organizations can have overlapping roles with market intermediary organizations. Both are specialized agents working closely with organic producers. However, market facilitating organizations are not part of the supply chain of organic products. For instance, the market facilitation role of Grupo Ecologica Peru has made feasible the Bioferia Miraflores since most small scale producers are not able to assume facilitating tasks because of lack of time, resources and specialized skills. Furthermore, it is difficult for small scale producers to be heard in policy making domains and in the media to advertise the Bioferia Miraflores. Another emblematic example of market facilitating organization is PIDECAFE, a NGO supporting the access of small scale producers associations to international markets.

Market intermediaries aim to support the commercial, technological and managerial side of organic production. Market intermediary organizations can buy organic food from small scale producers to supply higher volumes to large companies. In this way, market intermediaries channel organic food supply from producers to buyers or consumers. Market intermediaries are necessaries in local and international supply chains. Market intermediaries organize quantitative data of commercialization and market dynamics for the decision making of their small scale providers, including information about productivity, productiveness, production infrastructure, market prices and forecasts. Market intermediaries also translate farming techniques to producers. Working closely with producer associations, market intermediaries answer to the local and international demands of organic products. An example of a market intermediary organization is Grupo Ecologica Peru, a NGO with a growing business division. Beside the facilitating role, Grupo Ecologica Peru is mediating the supply of organic products from small scale producers to supermarkets under the brand 'Ecologica Peru'. The market intermediary role implies that Grupo Ecologica Peru supervises the whole chain of supply. Moreover, Grupo Ecologica Peru transfers its experience in organizing production and commercialization strategies to its providers, so it can be able to supply directly to buyers for supermarkets, hotels and so on (Pardo, 2008, interview).

Agents in the market access discourse are also technocratic NGOs. They hold specialized knowledge and expertise on organic farming techniques. Although technocratic NGOs do not have expertise on market, their knowledge on production is market oriented. An example of a technocratic NGO in organic farming oriented to markets is Huayuna.

Through their representative organizations and platforms, organic producers have agency. Despites producers have very diverse backgrounds, skills and formal education, they are capable and 'brave' with a strong will, ambitious and boldness. In a few cases small scale producers can even be authoritarian with their fellows. The main motivation of organic producers subscribing to this discourse is profit making and to conquer competitive markets. Discourse subscribers claim that the image of 'poor' farmers and 'sacrificed' ecologists is a marketing strategy of organic producers to captivate consumers. In the words of Pardo (2008, interview) 'organic producers hide their main motivation, which is profit making'. Organic producers seek to change liberal market conditions in order to be treated fairly in accessing it and benefiting from its advantages. A few examples of organic enterprises that fulfill this discourse are the producer associations of Matucana and Biofrute, and the enterprises La Cabrita and Vacas Felices. The first two producer associations are affiliated to ANPE. The last two producer associations are affiliated to Grupo Ecologica Peru. Most organic enterprises subscribing to this discourse have the vision to supply supermarkets and to export their products to international markets. This is the case of APPEAP that is exporting mango, the producer association of Santa Cruz valley of Satipo that is exporting organic coffee, the producer association of Ucayali that is exporting camu camu and the producer associations of Puno that is exporting caniwa powder (Trejo, 2008, interview). Other examples are the producer associations CEPICAFE and APROMALPI (Paz, 2008, interview). Commercialization towards competitive local and foreign supermarkets are seen by small scale enterprises as a sort of social status, self-stem and citizenship.

The identity of producers as business people in the market is in a formation stage. Although small scale producers run their own businesses, most of them do not prefer to be called 'businessmen'. For most producers it is still difficult to build an identity as business people. Actually, according to the producers subscribing to this discourse, business people refer to the owners of larger companies and they are not well seen by producers. Similarly organic producers that transform food are usually called 'processors', rather than small agro-industries.

Conventional NGOs in this discourse are ineffective agents. Conventional NGOs are criticized in the sense that during 20 years of organic production producers have received only 'superficial' information of organic farming technologies, but nothing about how to handle organic techniques and managerial skills to produce for competitive markets (Pardo, 2008, interview). Supermarkets and other profitable markets have not been well explored yet by agro-ecological NGOs, as most of them focus on social concerns of organic agriculture. Most agro-ecological NGOs are busy with ideological discussions whether or not producers should move towards competitive markets. Conventional NGOs have a lack of knowledge and experience on market and business skills. Most producers do not have expertise in commercialization (Trejo, 2008, interview). The discourse subscribers point out that actually most conventional NGOs are biased towards opposing globalization and resisting liberal markets. Indeed, the need to access competitive markets has been raised by small scale producers themselves. Organic producers claim that they, not the NGOs, raised concerns about where to sell and how to make consumers pay higher prices for their products. Conventional NGOs are referred to as 'blind' organizations since they do not have sufficient expertise to cope with the needs of producers in competitive markets. Furthermore, discourse proponents claim that most agro-ecological NGOs in Peru do not have a clear position about the role of NGOs in markets. Conventional NGOs are considered naïve because their capacity building strategies are not enough for small scale producers to access competitive markets. In this context, the market access discourse question traditional ways of supporting small scale producers.

Organic consumers have agency. Organic consumers realize the power they have in choosing what they want to eat and what they do not want. They have individualist motivations regarding their health and nutrition. Organic consumers do not care much about social and environmental benefits of consuming organic food. Overall organic consumers in this discourse are people in search for a more individual holistic life-style. Organic consumers are usually sports people, elderly people, householders, pregnant women and spiritual people. The profile of organic consumers is people with higher education and middle to high income. Organic consumers have performed important roles in providing feedback to producers, including packaging design and ideas of new products. The feedback has allowed producers to growth in production. As Pardo (2008, interview) stated 'nowadays more a more consumers ask for food of high quality, and look at the technical specifications and ingredients of the products. Ten years ago the attitude of consumers used to be a bit more passive'.

Organic production in rural areas might diminish migration and improve the economy of family producers. Nowadays people are aware of environmental problems such as climate change but most people are not aware of the impact of their life-style on environmental degradation. Consumers have to play a more active role in protecting the environment, sharing interests between consumers and producers. Shared interest is understood in terms of: 'each one gets what one wants'. It means that producers get income and consumers get healthy food. Indeed, according to the discourse proponents, the search for common interests among organic producers and consumers might be the basis for elaborating strategies of environmental protection (Pardo, 2008, interview).

Other agents are organic certification bodies. These are companies that grant third party certification to endorse the organic condition of the products at the market. Certification bodies also help to convert production from conventional to organic. Certification bodies play relevant roles in implementing certification schemes for groups of small scale producers such as the internal control system (ICS). ICS is specially a requirement for organic producers providing their production to international markets. Proponents of the discourse put in doubt the feasibility of the Participatory Guaranty Systems (PGS) as certification scheme for the market.

The national government is seen as an incapable agent. The market access discourse proponents claim that the national government does not realize the value of organic production as an opportunity for rural development and poverty fighting. The support of the national government is considered not enough to take advantage of the opportunities that international markets offer for small scale producers. Despite fact that the national government has taken some initiatives in building a national institutional framework for organic production, a gap is seen between policy and practice. The national government, according to the market access discourse adherents, sees NGOs as being 'against' development. To say so, discourse subscribers refer to the statement by the former President of Peru in an article titled 'El Perro Del Hortelano'. In the words of Pardo (2008, interview) 'promoting ecological agriculture for small scale producers, developing local organic market and advocating fair conditions to access competitive international markets does not mean to be 'against' the policies of national development'. Although national government has the responsibility to protect natural resources and establish proper conditions for organic production, it is seen as favoring agrochemicals and black-box technologies that make producers dependent. Furthermore, the governmental aid programs are seen as initiatives that have eroded the capacity of producers in rural areas to produce their own food.

Key metaphors and their rhetorical devices

The market access discourse has several metaphors. The first one is the notion of 'translators' to refer to facilitating and intermediary NGOs decoding complex market information for small scale producers.

Another favorite metaphor is the perception of business as something 'benign', not necessarily as 'something bad'. This perception is different from that raised by the proponents of the market adaptation discourse. Business is a source of learning for both NGOs and producers to access market in a good position. NGOs should learn about business standards and quantitative measurements. An example is Grupo Ecologica Peru that is learning business approaches and searching a proper organizational model to support producers to market access. In this discourse market goes in line with social concerns. Quoting Pardo (2008, interview) 'social challenges can be undertaken through business and market, especially 'the war' against poverty. The market is not necessarily divorced from social concerns'. Market NGOs are gaining expertise on business rationales with a social component. The perspective of seeing competitive markets with fear is no longer valid in this discourse.

Other metaphors refer to markets as a 'supernatural force' or sort of 'god' that people cannot control. Discourse proponents claim that the market is a global process beyond peoples will and that nobody can stop. Discourse subscribers argue that, for instance, nobody has the power to stop the expansion of supermarkets in Peru. In the eye of this discourse, efforts made against liberal market are in vain. Not to be connected to the global market means to be marginal and excluded.

The liberal market economy is considered a 'power' that people should 'surrender' to. The surrendering is unavoidable as the liberal market economy is the 'real life' and 'real world'. The metaphors describe the fact that economic rationality rules modern life. 'Real life' and 'real world' refers to all things such as the individualism, self-interest and selfishness, against nature, the need of money and other material needs. According to this discourse, organic production has to be economically feasibility to be sustainable. The connotation of both rhetorical devices is negative in terms that they control the human wishes and wills.

In referring to the national government, the metaphor of 'incapable entity' is usually used.

5.3.4 Discourse of market democratization

Food sovereignty first!

Small scale producers produce for self-consumption and the surplus is commercialized through local markets, especially at farmers' markets in cities and towns. By doing so, the organic market is democratized to wider low and middle income consumers rather than keeping it as privilege for high income consumers. Small scale producers construct their own institutional infrastructures to connect with institutions of the liberal market economy and lobby against the national government to obtain public policies and legal frameworks on organic production. In the long-term organic producers seek to reach competitive markets *vis-à-vis* transforming the liberal market economy. Table 5.8 provides an overview of the analysis of the discourse of market democratization.

Basic entities acknowledged

The liberal market economy is acknowledged as biased for profit making and missing social and community considerations. Organic producers subscribing to this discourse claim that the liberal market does not work for small scale producers and it undermines community values. Social strengthening of communities is a central condition in the discourse to create a fair relationship with liberal market institutions. Even though the liberal market is acknowledged, small scale organic producers seek the transformation of its basic foundation in the long-term. Proponents claim that it can be achieved through strengthening shorter chains of supply rather than the usual model of global supply chains. Following the former, small scale producers seek to be much closer to consumers, leaving away several intermediaries.

Organic markets are understood in terms of food self-sufficiency. The local production of food for self-sufficiency and the country's food sovereignty are the core claims of the market democratization discourse. According to Silverio Trejo (2010, interview), the former president of ANPE and owner of APICEL, a small organic enterprise, the priority of ANPE is to generate economic development for family farmers in rural areas. This means enhancing the commercialization of organic products at local level. Market democratization means make organic food available to all consumers, regardless of their economic condition. Discourse proponents claim that small scale producers should depend mainly of what they produce at their own organic farms. Trejo (2010,

Table 5.8. Discourse analysis of market democratization.

Elements	Market democratization
Basic entities recognized or constructed	Liberal market
	Organic market is understood peculiarly and limited to local
	Globalization
	Participatory Guaranty Systems (PGS)
	Organic consumers
	Sustainability
	Conventional NGOs
	Government
	International cooperation agencies
Assumptions about natural in relationships	Sense of community and shared cultural roots
	Struggle against hierarchy
	Human society depends on nature
Agents and their motives	Small scale organic producers: enlightened, sensible and imaginative
	Selfish larger business people
	Self-interested conventional NGOs
	Guarantors of PGS
	Consumers with scarce awareness about organic food
	Careless and unfair national government
Key metaphors and other rhetorical devices	Emancipation
	Harmony with nature
	Struggle among good boys and bad boys. Most agents are conspirators against producers
	Market is dirty, bad and threatening
	Dissidents: deserters of the principles of agro-ecology

interview), states 'the day that I have money I eat like poor people and the day that I do not have it I eat like rich people'. Export organic food to foreign markets has no priority. The priority is the development of local organic markets. Although the liberal market is accepted, it does not play a central role in the discourse. Putting markets first means for discourse subscribers to produce organic food only for selling and buying 'conventional' food for their own family consumption.

The global arena is acknowledged and it is seen as a new playground for NGOs, small scale producers, large producers and the national government to collaborate, bargain, struggle, negotiate and gain better positions. Small scale producers are connected with fellow organizations at regional and global levels in order to increase their power against the national government. For instance, ANPE intends to influence national agricultural policies in favor of organic agriculture through influencing IFOAM and FAO. As Trejo (2010, interview) stated 'this strategy is the best way to influence the national government since it obeys global policies. For instance, the standard of third party certification was born at IFOAM, not at the national governments. National governments just implement what international agencies establish'.

The market democratization discourse acknowledges the Participatory Guaranty Systems (PGS) as the scheme that assures the organic condition of food at the market. PGS matches the need of guarantee the organic condition and the need to socially strengthen local communities. PGS is seen as a fundamental strategy to develop organic markets for low and middle income consumers, and strengthen social cohesion of producers and their communities. PGS generates trust and a sense of community among producers and it is appropriate for producers commercializing at farmers' markets located in small towns and cities. Actually, PGS is the guaranty scheme of organic production at 8 Bioferias promoted by ANPE in Peru. According to Trejo (2010, interview), small scale producers should be able to export with PGS certification on the Latin American continent. Even more, he says, 'producers at the Intercontinental Network of Organic Farmers Organizations (INOFO) are suggesting that PGS should guarantee the trading of organic products by small scale producers at the global level'. Discourse proponents claim that third party certification is not feasible for most of the small scale producers due to its costs, the involvement of big companies and the generation of the 'surplus' for organic products. PGS widens the consumption of organic products compared to organic certification, as the high costs of third party certification is transferred into high prices of organic products.

Sustainability is constructed based on the life experiences and the cultural values of small scale producers. In the market democratization discourse, organizational structures of producers and the agro-ecosystems are the core components of sustainability. Agro-ecosystems are seen as the space where wild nature and crops interact in harmony. In the eyes of the discourse subscribers, technology development has to be controlled since it impacts human beings. Transgenic crops are seen as something that erodes biodiversity and the economy of small scale producers. The discourse highlights the value of research for the genetic improvement of crops but in a respectful way towards nature.

Assumptions about natural relationships

The central relationship assumed by the market democratization discourse is the strong sense of community and shared cultural roots. Actors endorsing the discourse are driven by values of trust and mutual learning. Discourse proponents do not speak about adapting or accessing the liberal market but about a fair relationship with market institutions. Social strengthening of communities is considered a central condition to create such fairness. There is a struggle of actors against hierarchy. Small scale producers, local NGOs and community based organizations struggle for emancipation, power and self-affirmation against the national government and agro-industries. In the eyes of this discourse, the struggle of small scale organic producers against larger 'conventional' producers and agro-industries takes place in a hierarchical world. However, powerful international actors, for instance IFOAM and FAO, can also ally with small scale producers against those opposing forces. In this struggle, larger producers and agro-industries base their power on economic assets, while the power of small scale producer power is based on their 'social' force.

Struggles not only take place against larger 'conventional' agro-industries but among producers themselves. The rise of organizations and platforms of small scale organic producer associations and enterprises has generated the unfolding of struggles at two levels; within and beyond organic producers. In the first case, the creation of ANPE has meant the rise of a new actor with its own perspective and with more autonomy from conventional NGOs. In the second case struggles between organizations of small scale organic producers and large scale producers means divergence in ideological positions. For instance, meanwhile ANPE is requesting the government to support ecological agriculture, the National Convention for Peruvian Agriculture (CONVEAGRO) requests subsidy for agrochemicals (Trejo, 2008, interview). However, it is also worth mentioning that at the two levels there is room for cooperation as long as actors share interests and values. Agro-ecological NGOs, non-organic producers and organic producers collaborate in different degrees through their platforms, especially to negotiate with the national government. Although opposing interests and values cause clashes between organic and non-organic producers, common concerns usually bring them together for cooperation. The current closer cooperation between ANPE, CNA and CONVEAGRO is a good example of the latter.

The relationship of human society with nature is one of dependency. People depend on nature to get high quality food. Nature as a whole has its own value in terms of social, cultural and ecological meaning. People do not have a hierarchical position on top of nature. Instead, people have the responsibility to protect nature through practicing non-harmful agriculture. Proponents of this discourse are implementing projects to protect forest and valleys in several sub-national regions of Peru, including Amazonas, Ucayali, San Martin, Madre de Dios, Ancash and Cusco in cooperation with municipalities (Trejo, 2008, interview).

Agents and their motives

Small scale organic producers are considered key agents in the market democratization discourse. Small scale organic producers also include small organic food industries and organic handy craft enterprises. They consider themselves the guardians of nature and the saviors of environmental disaster. In their eyes, organic producers are the exemplary actors with environmental consciousness, self-steam and responsibility. Being able to emancipate from conventional NGOs, especially in financial terms, organic producers have shown strength in facing new challenges. Motivations of producers have to do with strengthening the capacities of communities to produce their own food. Small scale organic producers through their platforms are responding with campaigns to catch public attention against transgenics, synthetic fertilizers and chemical pest control. In the words of Pardo (2008, interview) 'we cannot stay with crossed hands, while large companies clash us'.

Small scale producers want to be part of a globalizing market but with their own values, assets, views and approaches. As small scale producers contest the logic of a liberal market economy, they imagine new market relations. Small scale producers do not oppose or resist the globalizing liberal market, but call for a fair treatment, and a just relationship between producers and liberal market actors. Organic producers accept the market, but giving priority to the social motivation of organic production. Small scale enterprises endorsing the discourse are looking for profit making in the long-term. Discourse proponents call for the access of producers to global markets in the same way that foreign products enter their countries. For example, currently small scale producers cannot freely trade their organic products beyond country borders. Discourse proponents claim that the use of PGS needs to be regulated by the national government. Given the fact that the discussion of PGS in trade agreements between Peru and Bolivia are not included by the national governments, ANPE together with Bolivian and regional actors are influencing both national governments (Trejo, 2008, interview). Small scale producers subscribing to this discourse intend free trade in South America using the PGS scheme and even export native crops-based organic products to developed world markets.

Small scale organic producers have also structured their own associations, NGOs and platforms to dialogue adequately with liberal market proponents and to lobby the national government for proper public policies on organic agriculture. For instance, the Agro-ecological Consortium is a new platform established by organic producers in cooperation with NGOs to dialogue with the national government.

Larger businesses have agency for selfish purposes. Discourse proponents claim that most large agro-exporters affiliated to the Peruvian Exporters Association (ADEX) are buying large extension of land. This is seen as a threat for small scale producers. Small scale producers are afraid that they might end up as contractor of larger business as governmental policies are designed to produce on large scale (Trejo, 2008, interview). In the discourse, larger businesses are based on individualistic values rather than communitarian values. In addition, proponents claim that social strengthening is what makes larger agro-exporters 'fear' small scale organic producers.

Conventional NGOs have agency but that is becoming less influential. Most NGOs are considered paternalistic, ineffective and self-interested agents. Producers and their organizations are considered central actors in organic production. Currently conventional NGOs are more interested in working with producer organizations, not other way. Small scale producers have now the power to accept or refuse cooperation of NGOs. Depending on whether cooperation is of benefit, NGOs are accepted as partners. Other NGOs are seen as self-interested agents that 'use' producers only to get funding from international cooperation agencies. Actually, the misuse of funding is seen by the discourse proponents as the cause of the crisis of most conventional NGOs. So-called assistencialism of most conventional NGOs is questioned in this discourse. As Trejo stated (2008, interview) 'producers should pay costs involved in production and commercialization by themselves; not by NGOs. Otherwise producers do not realize the value of things, as what is for free is not sustainable'.

Other agents in the market democratization discourse are the guarantors of the PGS scheme. Guarantors are producers and consumers that assure the organic process of food production. The direct involvement between producers and consumers makes PGS less onerous in terms of paperwork and record-keeping requirements. In contrast, in third party organic certification schemes everything is handled by the certifier company and most small scale producers are not able to pay the high costs involved. Certification companies and PGS guarantors are based on different rationalities. Farms that produce organic crops in monoculture can be awarded with third party certification but not with PGS, claim the discourse proponents. Discourse proponents also claim that a farm might be granted with third party organic certification, though its owner and his/her family is not well feed. It is possible to meet third party certification without putting the social concern at the core of the business. Additionally, the discourse proponents claim that third party organic certification does not automatically drive small scale producers to become sustainable enterprises, as it is not a proof of social sustainability of producers.

Consumers are people with agency, but with scarce awareness about organic food. Proponents agree that consumers are not enough informed about what organic food really means. In the words of Pardo (2008, interview) 'consumers have been told that organic food has to be more costly. This is true if the product has an organic certification. But, if the product has a PGS, it is affordable for most consumers'. In the eyes of the discourse subscribers, PGS enhances the direct interaction between producers and consumers adding social value to organic consumption.

The national government has agency to act but it is careless and unfair with small scale producers. The national government puts obstacles and ignores organic producers. Instead, the national government provides all support to agro- industries and agro-exporters. In prioritizing agro-exports, the government undermines development of small scale producers and the local market. Discourse proponents claim that the national government works in favor of big capital interests, not of small scale producers. They claim that the national government does not recognize the social services of small scale producers to society and it has almost abandoned rural development. As Pardo stated (2008, interview) 'organic production is not the priority for the national government. The governmental policies are oriented to agro-exporting despite most part of food consumed in urban areas is supplied by small and medium-sized producers. While politicians are busy in lowering customs tariffs to allow the entering of imported food, there are no incentives for local small scale organic producers'. The support to organic production depends of the goodwill of individual policy makers. In the eye of the discourse proponents, the low institutionalization of organic production in governmental agencies shows the low interest by the national government. For instance, ANPE's lobby strategies are based on personal connections to influence the governmental agencies SENASA, MINCETUR, PRODUCE and MINAG. Small scale producers subscribing to this discourse argue that in the eyes of the national government, they are 'The Dog of the Orchard'. In the words of Trejo (2008, interview) 'the government ignores that The Dog of the Orchard produce most of the food that is consumed in Peru: 80% of products comes from small scale and medium-sized producers. As small scale producers pay all the costs of production and commercialization, small scale producers subsidize to consumers located in the cities, especially in Lima. This causes people in the rural areas to be poor because the subsidy comes from small scale producers and goes to the national government and urban consumers. As a result, rural areas are getting empty because of migration to large cities'.

Another example of the unfair attitude of the national government, according to the market democratization discourse proponents, is its unclear position regarding the PGS scheme. The government does not support PGS as a valid system to assure the organic condition of organic products at the Peruvian market. SENASA, the national authority in organic production, does not want to recognize PGS as a valid certification scheme to commercialize organic food legally in Peru. According to the law N°29196 'Promotion of Organic Agriculture', approved in 2008, to be labeled as organic food, it has to be certified by a certifying company authorized by SENASA. That law also mentions the PGS as a valid scheme to commercialize organic food for the internal market. However, producers using PGS cannot use the label of 'organic' because of the lack of a regulation of the aforementioned law. According to the discourse proponents, SENASA is in power to recognize PGS as a valid assurance scheme. In the words of Trejo (2008, interview) 'this lack of interest of the national government has to do with interests from organic exporters that do not want that small scale producers commercialize their products using PGS massively at the local market. Companies tied to third party certification are afraid that PGS expands to local markets'.

Key metaphors and their rhetorical devices

The market democratization discourse is rich in metaphors. 'Emancipation' is the common metaphor. It refers to the capability to break dependency relationships, control and domination. The founding of organizations in which producers have decision power is seen as prove of a more autonomous character of small scale producers. First of all, small scale producers feel emancipated from dependency and control of conventional NGOs, especially in financial terms. The motivation of promoting self-consumption and food sovereignty is also an extension of the emancipating profile of small scale producers. In this sense, producers see themselves as heroic in their emancipating endeavor.

Another metaphor is 'harmony' to refer to the relationship with nature. Harmony is understood in terms of life-style, consistency, respectful attitudes against nature and agro-ecosystems. According to the market democratization discourse, without a harmonious relationship with nature sustainability cannot be reached.

'Struggle' is also a common metaphor in the discourse that highlights a worldview of fighting. The struggle is one of social versus economic concerns, and of small producers versus larger agro-industrial perspectives. Small scale producers defend themselves against larger scale producers and larger agro-industries. Small scale producers use their social power as a weapon against the economic power of large agro-business. As the national government, larger 'conventional' agro-industries and larger 'conventional' producers are believed to be against small scale producers, proponents of the discourse endorse the idea of 'conspiracy' against small scale organic producers.

Other metaphors refer to the market. The liberal market has the connotation of something 'bad', 'dirty', 'threatening' and 'contaminating'. Proponents claim that liberal markets make social and environmental concerns less important. This means, for discourse subscribers, that economic concerns are a top priority and social concerns are dropped. However, markets are also seen as opportunities for small scale producers in the long-term. Markets are triggering the formation of global movements of small scale producers, INOFO being an example.

Other metaphors are related with views about the world. The world is seen as 'white or black'. The struggle in the world is ultimately between 'good guys' and 'bad guys'. Bad guys are actors that are selfish, disrespectful with nature and slaveries of money. Good guys are safeguards of the social perspective of organic food at the market. Nestle, Coca-Cola and other transnational companies are not well perceived. Supermarkets are also seen as a privilege for rich people. It is quite common in the discourse to use the metaphor of 'dissidents' to refer to the small scale organic producers that abandon the principles of agro-ecology and social justice, and embrace the liberal market (Trejo, 2008, interview). Producers endorsing the liberal market are considered inconsistent in their arguments. Quoting Trejo (2008, interview) 'conventional producers and dissidents claim to be ecological, but they consume Coca-Cola and other conventional products in their daily life, and usually they buy at supermarkets, not at farmers' markets'.

5.3.5 Conclusions

a. Based on empirical research of the organic production in Peru, three discourses could be indentified: the market adaptation discourse, the market access discourse and the market democratization discourse. Each discourse encompasses a particular storyline, showing that organic production is not a unique body of thought but an arena for diverse perspectives and debates. The market adaptation discourse highlights the issue that actors are pushed for change by powerful agents. This means that especially NGOs and small scale producers are forced to obtain new capacities and be well equipped to adapt to the liberal market economy. Assisting small scale producers adapt to liberal market becomes a new role of NGOs, as failing to adapt means marginalization and exclusion. The market access discourse underlines the issue that actors are eager to move to competitive markets and that small scale producers play a central role in this ndeavour by becoming suppliers or intermediaries to competitive markets. Small scale producers are determined to conquer profitable markets. The market democratization discourse emphasizes the fact that small scale producers primarily produce for self-consumption and the surplus is commercialized through local markets in cities and towns. According to this discourse, organic food is oriented to wider low and middle income consumers rather than as a privilege for high income consumers.

b. All discourses acknowledged the central role of the liberal market but it is approached from three perspectives: 'adaptation', 'access' or 'democratization'. The market adaptation discourse emphasizes the creation of small models of commercialization, the market access discourse emphasizes profitable local and international markets, and the market democratization discourse emphasizes mass consumption of organic products. The different understanding of the market in each case means also different relationships with nature. The market adaptation discourse recognizes nature as the guarantor of human survival whereas the market access discourse has hierarchical, individualist and anthropocentric views on nature. In the case of the market democratization discourse, it endorses the dependency of human society on nature. The market democratization discourse views on nature are closer to those of the market adaptation discourse than those of the market access discourse.

c. Despite the fact that each discourse embodies a particular storyline, the three discourses have commonalities. The three discourses encompass the liberal market as a playground for small scale producers, criticize the national government and consider organic consumers as allied actors. While in the market adaptation discourse NGOs are the central agents, in the market access discourse and the market democratization discourse small scale producers and small scale enterprises play central roles.

d. Organic production discourses are rich in metaphors and rhetorical devices. Metaphors in the discourses refer to NGOs, producers, business, markets, sustainability and the national government. Differences and similarities are found in the images of those entities given by the metaphors. However, metaphors coincide in showing that the discourses of organic production are embedded in perspectives of hierarchy, prejudges and struggles.

5.4 Discourses of organic production in a two-dimensional policy realm

Based on the discourse analysis of organic production a two dimensional axis can be constructed to allocate the three discourses on organic production (Figure 5.2). The two dimensional axis grasp the social (axis A) and economic (axis B) realm. On the vertical axis (A) the focus is on the social foundation for structuring society, whereas the horizontal axis (B) concentrates on the economic foundation for organizing society.

The extreme poles of axis A represent 'individualistic' and 'communitarian' values. 'Individualistic' values highlight individual interests over collective interests and a faith in the capacity of individual action and ambition to create wealth and to bring about progress, whereas the 'communitarian' values highlight collective values and common identity of people within a shared geographical scope. In the first case, individual interests are the base for relationship among actors. Actors seek for single or group benefits rather than collective welfare of the whole society. In the second case, what is central is the social strengthening of local communities. The collective values are, among others, solidarity towards people in need, a sense of community and cultural identity.

The extreme poles of the axis B represent 'self-sustaining' and 'liberal market' values. 'Self-sustaining' values highlights the position of not requiring any outside aid, support, or interaction for autonomous survival, whereas 'liberal market' highlights the mainstream liberal market economy where business has a central role in production and commercialization of goods and services at the global level. In the first case, organic production is mainly oriented for self-consumption of the family, the community or the country under the frame of food sovereignty. In this case the

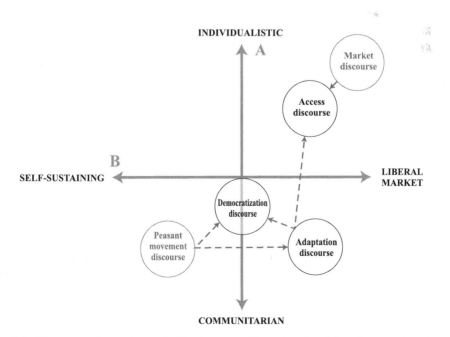

Figure 5.2. Discourses of organic production in a two-dimensional social and economic axis.

economic structuring of society does not depend on external provisioning or exports of goods and services, including food. In the second case, giving the worldwide spreading of the liberal market and globalization, actors are interlinked globally. The connectedness of actors form local to global scale and the mobilization of goods and services, including food, in a global scale are the basis for the economic structuring of society.

Based on their storylines, the organic production discourses are allocated in the aforementioned two-dimensional matrix (Figure 5.2). The market adaptation discourse is allocated in the crossing of 'communitarian' and 'liberal market' values. It means between the central role of business for economic structuring of society and the social strengthening of local communities. The market access discourse is allocated in the crossing of 'individualistic' and 'liberal market'. It means between the interest and action of individuals to create wealth and to bring about progress and the central role of business in production and commercialization of goods and services at the global level. The market democratization discourse roots strongly on 'communitarian' values and it matches 'self-sustaining' and 'liberal market' values. It locates between the autonomous supply for self-consumption and the social strengthening of local communities. The organic production discourses have evolved from the 'peasant movement' discourse. The peasant movement discourse is allocated in the crossing of 'self-sustaining' and 'communitarian' values. Particularly the market adaptation discourse has its origin in the peasant movement discourse. The market democratization discourse has evolved from the market adaptation discourse but it has been influence by the peasant movement discourse as well. The market access discourse has emerged from the market adaptation discourse. The market access discourse diverges strongly from the other discourses. The position of the market adaptation discourse and the market democratization discourse are quite close to each other in the two-dimensional axis, despite differences in their views.

5.5 Major patterns and trends: networks and discourses

In this section networks and discourses are put together in a single framework (Figure 5.3) to understand the main patterns and trends resulting from their relationships.

The main patterns and trends identified from the relationships between networks and discourses of organic production are the followings: (1) the rise of new discourses; (2) the rise of new actors; (3) the embedding of discourses and networks in a liberal market; (4) changes in the policies of international cooperation agencies; (5) closer interaction between NGOs and small scale enterprises; and (6) a tension in the discourses between organic production and environmental protection.

First, new discourses of organic production have emerged during the last two decades, and NGOs are their main proponents. Over time NGOs changed their perspectives and endorsed new discourses. During the 1970s and 1980s most NGOs used to subscribe to the peasant movement discourse, only promoting self-sufficiency of farmers. In the 1980s in their efforts to assisting small scale farmers to commercialize their products at local farmer markets meanwhile protecting agro-ecosystems, some NGOs subscribed to the discourse of market democratization. During the 1990s NGOs started to move from solidarity to market values, endorsing the discourse of market access. Other NGOs called for adapting to the globalizing liberal market, subscribing the market adaptation discourse. The mainstreaming of the market access discourse *vis-à-vis*

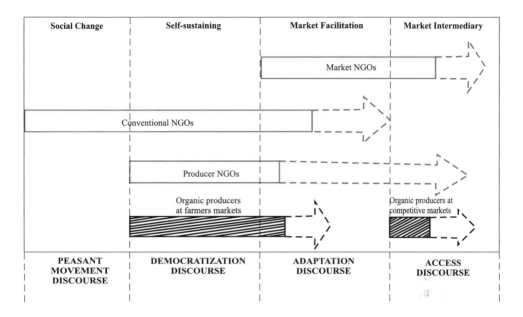

Figure 5.3. Main discourse and network patterns and trends of NGOs, organic producers and organic enterprises.

the expansion of market NGOs and small scale organic enterprises supplying larger buyers is a clear tendency in the networks of organic production. Similarly, organic farmers' markets will likely keep expanding in number *vis-à-vis* the further organizational strengthening of producer organizations. As a result, new roles of agro-ecological NGOs have emerged, such as market facilitation and market intermediation, triggered by the emergence of new discourses of organic production. The trend is that while some NGOs will keep evolving in facilitating markets others will become truly market intermediaries. Even certain NGOs are expected to become consultancy-like organizations. However, I would claim that neither the peasant movement discourse catches the actual interests of agro-ecological NGOs and SMEs nor that the new views mean that NGOs have totally abandoning their former views.

Second, the rise of new discourses of organic production is materialized in the emergence of new actors. New actors include new types of agro-ecological NGOs and small scale organic enterprises. New types of agro-ecological NGOs are market NGOs and producer NGOs. These new type of NGOs have emerged to perform new roles and fulfill new market needs of small scale producers. Most producer NGOs focus on self-sufficiency strategies and market facilitation roles endorsing the market democratization discourse. ANPE is a key example. However, a few producer NGOs are moving to market intermediary roles, subscribing to the market adaptation discourse. The Bioferias are examples. Producer NGOs even might endorse the market access discourse, as shown by CEPICAFE, APROMALPI and CEPIBO. Producer NGOs appear because of the need of producers to harmonize self-provisioning of food and the commercialization of their production at local markets. Producer NGOs are gaining leadership positions at local and global platforms of organic production. Market NGOs started as market facilitators but they are now moving to market intermediary roles. It means market NGOs endorse either the market adaptation discourse or the market access discourse. Market NGOs emerge due to the need of specialized agents to facilitate and intermediate the access of small scale producers to competitive markets. Examples of market NGOs are Grupo Ecologica Peru, El Taller and PIDECAFE. Although conventional NGOs are not a new type of NGOs, they are also changing their roles. An example of conventional NGOs is the Centro IDEAS. Increasingly, conventional NGOs are becoming market facilitators. Conventional NGOs will not easily become market intermediaries but they will set up their own businesses, for instance organic shops and organic food restaurants. Small scale organic producers and enterprises also emerge as legitimate actors following the rise of the market democratization discourse and the market access discourse. This emergence also has to do with the self-organizing capacity of producers and the presence of a 'niche' market for organic food at national and international levels. The emergence of new actors in the networks has also triggered disputes for power. It is hard to know to what extent the rearrangement of power in the organic production networks between conventional NGOs, producer NGOs and market NGOs will also contribute to mainstreaming the 'organic' perspective, considering that the conflicts between them are weakening the Peruvian agro-ecological movement.

Third, the central position of the liberal market in the world and in the national economy is shaping the discourses of organic production. The liberal market is approached by agro-ecological NGOs from three different perspectives; adaptation, access and democratization. Moreover, the liberal market is motivating producers to organize themselves in associations and cooperatives, diversifying their products, and even combining the commercialization of organic and non-organic products. For example, CEPICAFE with the support of the NGO PIDECAFE supplies organic coffee, fair trade coffee and non-organic sugar cane for international markets. The mainstreaming of the liberal market does 'pressure' actors not only in terms of funding availability but also in terms of considering access to competitive markets as the only way to improve conditions of small enterprises. However, this embedding of the liberal market in the discourse of organic production does not mean that the views of advocacy, influence and checking are fully abandoned.

Fourth, NGOs depending of funding from international cooperation agencies are being forced to change their priorities and approaches. Depending on the type of NGOs, they are looking for different survival strategies. In the case of conventional NGOs, they are moving from traditional charity approaches to market facilitation roles. In the case of market NGOs, they are deepening their market intermediary roles and keeping their market facilitating roles. Producer NGOs are also deepening their market facilitation roles. Producer NGOs plan to set up business branches, for instance organic stores, consultancy companies and agro-eco-tourism resorts. Under this new scenario of changing policies most agro-ecological NGOs will be impacted in way to finance their supportive activities towards organic producers and enterprises. If agro-ecological NGOs, especially conventional NGOs, do not look for new funding and generate new income, they will likely disappear.

Fifth, the closer interaction between NGOs and small scale enterprises includes business perspectives in the traditional social perspectives of agro-ecological NGOs. As a result of this process, 'hybrid' NGOs are emerging. Hybrid NGOs are business-like NGOs that increasingly build up ties with market actors and subscribe to liberal market perspectives, but keep a strong social development component in the discourse. Hybrid NGOs might remain non-profit or become for-profit. In the former case, NGOs become shareholders of business or establish a business. In the latter case, NGOs establish a division within their organization that deals with businesses. Hybrid NGOs are promising opportunities. But the case study shows also that they are challenging. The experience of Grupo Ecologica Peru shows that harmonizing actors and interests in a single organization is hard. The business component and the NGO component of Grupo Ecologica Peru are not harmonized yet. In the words of Pardo 'the components are like two persons in a room but they are not hugging each other yet'. Harmonizing means to define a proper organizational model to handle business by answering quickly to market demands, to strengthen capabilities of the organization and suppliers, to develop the demand of potential consumers through social marketing and to raise awareness to increase demand.

Sixth, a tension between organic production and environmental protection can be identified in the discourses. The tension has to do with the fact that in environmental protection nature is the target, but in organic production producers are at the centre of interventions. In competitive markets the tension between environmental protection and organic production is higher, especially when the market is far away from the place of production. But with, self-consumption this tension is less prevalent. All discourses of organic production do not take in account the potential environmental impacts of the growth of the national organic market, the need to apply cleaner technologies and the importance of efficient use of energy in production. Therefore, there is a conflict between the need of develop organic markets in Peru and the potential environmental impact of that development. In the words of Trejo (2010, interview) 'we plan to buy at least three tracks to transport organic products to distribute organic food to several farmers' markets at national level in Peru'.

5.6 Conclusions

a. New roles of agro-ecological NGOs are in line with the concerns of small scale organic producers and small scale organic enterprises. Currently, agro-ecological NGOs are performing three new roles: first, facilitating the development of organic farmers' market at the national level; second, intermediating the access of small scale organic producers to competitive markets at national and international levels; and third, building business skill capacities for small scale organic producers and enterprises. The emergence of these new roles means not only the diversification of roles but also the diversification of types of NGOs. Next to conventional NGOs, producer NGOs and market NGOs have emerged.

b. The organic production discourses are embedded by liberal market perspectives and by social and environmental concerns. The liberal market is acknowledged in all discourses of organic production, though it is approached from three different standpoints: 'adjustment', 'access' and 'democratization'. Therefore, the core adversarial viewpoint of the peasant movement has been overcome in the organic production discourses. The mainstreaming of liberal market perspectives in the networks implies that actors are more motivated by individualistic values. However, evidence shows that both communitarian and individualistic values are necessary to make agro-ecological NGOs and their networks sustainable. Social and environmental issues are also approached from divergent stand-points. While the environment is approached from ecocentric or anthropocentric points of views, social concerns are approached from communitarian or individualistic values. These differences generate a tension in the discourses between organic production and environmental protection, especially as cleaner technologies and the efficient use of energy in production are not considered in the growth of the organic market.

c. Cooperation contributes to the expansion of the organic production networks. Although conflicts and mistrust happen between producers, conventional NGOs and the new types of NGOs, and between agro-ecological NGOs and the national government, cohesion is stronger in the organic production networks. Differences in the discourses are the main factors explaining conflicts. However, commonalities bring actors (international NGOs, small scale organic enterprises, and consultancy and certifier companies) together against their common 'enemies': the transgenic and agrochemical companies, and their allies in state and markets.

d. Small scale organic enterprises challenge the power position of conventional NGOs and international cooperation agencies. Although the last two organizations remain influential in the organic production networks, organic producers and organic enterprises are gaining power. According to this case study, the increase in power is mainly due to three factors: the economic and social interlinkage of organic production, the networking capacity of small scale organic enterprises, and the attachment of small scale organic enterprises to local social and economic networks of cooperation.

Chapter 6.
Business social responsibility

6.1 Introduction

Business social responsibility, also known as corporate social responsibility or corporate citizenship, is a self-regulatory strategy of companies to legitimate the role of business in society and in the market economy. Business social responsibility implies that companies improve their social and environmental performance and gain trust of their stakeholders, while maintaining or enhancing their business competitiveness. Business social responsibility implies ethical, philosophical and operational considerations which are applied to manage and measure the social and environmental performance of companies, in four areas relevant to company: marketplace, workplace, community and environment (Cici & Ranghieri, 2008). The instruments used for such management and measurement include among others the balanced scorecard, reports of sustainability, and standards and ethical codes. Examples of standards and ethical codes are among others the SA 8000, the ISO 26000, the OECD guidelines and the Dow Jones Sustainable Indexes.

Mostly large companies have taken the lead in business social responsibility, but the principle has been extended to other organizations as well. SMEs, national governments and NGOs are endorsing social responsibility as ethical consideration to be accountable for their social and environmental impacts. The adoption of business social principles in SMEs is seen helpful to overcome, for instance, low quality employment, poor occupational health, substandard working conditions, violations of labor rights, lack of waste management, high energy consumption and unsustainable production. SMEs in Peru show mixed motivations in adopting social responsibility principles. While medium-sized enterprises mainly implement social responsibility actions regarding environment and workplace areas, micro and small scale enterprises are more focused on the community level. Regarding the areas of business social responsibility, SMEs in Peru perceive the market and community as very important for improving their economic development, while the workplace and environment are not considered very relevant. Overall Peruvian SMEs do not perceive social and environmental responsibility perspectives as relevant for their operations (Cici & Ranghieri, 2008). Usually the support of small business has been oriented at business growth in terms of only profit making, leaving other factors aside such as working conditions, occupational health and safely. Cici and Ranghieri (2008) conclude that there is lack of conditions that encourage SMEs to adopt social responsibility principles. According to Corral (2005) one of the critical obstacles for the adoption of business social responsible principles by SMEs is the limited awareness of the concept. Despite the aforementioned constrains, SMEs, in the form of individual enterprises or organized enterprises in clusters and associations, are part of value chains and constitute a good opportunity to apply business social responsibility (Villarán, 2004).

Business social responsibility involves in Peru a variety of NGOs. NGOs with a long history – which are called 'conventional' NGOs for the purpose of this research –, are part of the networks and discourses promoting business social responsibility. In this research conventional NGOs refer to NGOs founded in the 1980s that currently perform as market facilitators bringing to SMEs basic competences in social and environmental responsibility. Examples of conventional NGOs are among others CEDAL and CEDEP. NGOs handling a specialized 'know-how' on business approach, which – I called 'business' NGOs for purposes of this research, are part of the networks and discourses promoting business social responsibility. Most business NGOs appeared in Peru around the 1990s onwards. The mainstreaming of the liberal market in the national economy and the need of specialized knowledge and information by larger domestic and international companies, and their value chains are factors contributing to the rise of business NGOs. Business NGOs work as consultancy organizations and provide services to larger companies and thier small scale suppliers on expertise and guidance to implement business social responsibility as strategy of business management. Examples of business NGOs in Peru are among others Peru 2021, Responde and Impulsa Peru.

The chapter is organized in six sections. Section 6.2 presents the network analysis of business social responsibility and describes in detail the main networks identified in Peru: the 'social justice' network and the 'business' network. Afterwards, the main patterns, challenges and trends of the networks, and the main conclusions of the network analysis are presented. Section 6.3 presents the discourse analysis of business social responsibility and describes in detail the main discourses identified in Peru: the 'business upgrading' discourse and the 'corporate responsibility' discourse. Subsequently, the main conclusions of the discourse analysis are presented. In Section 6.4 the two discourses on business social responsibility are allocated in a two dimensional policy realm. Section 6.5 presents the major patterns and trends of the networks and the discourses of business social responsibility. Finally, the main conclusions of the chapter are presented in Section 6.6.

6.2 Network analysis of business social responsibility

The analysis focuses on identifying the key NGOs involved in the networks promoting business social responsibility for SMEs, the relationships among network members in terms of cooperation, resource exchange and power relationships, and the main network changes and challenges. The analysis concentrates on the network promoting business social responsibility for SMEs in Peru and their connections with other branches at regional and global level.

The social justice network and the business network are the two main networks that promote business social responsibility for SMEs in Peru. I have labeled the first network social justice network because the core issue in this network is promoting social justice. In this network, actors consider business social responsibility from a social perspective. The second network is labeled business network because the core issue in this network is profit making. In this network, actors consider business social responsibility from a business perspective. Each network includes a particular set of local, regional and global actors.

The social justice network is a network integrated by platforms and individual NGOs at local, regional and global level. Red Puentes Internacional is the Latin American platform of NGOs monitoring social responsibility of larger companies and promoting social responsibility in SMEs. Red Puentes Internacional involves NGOs, SME beneficiaries and other like-minded actors tied by cooperation, resource exchange and power relationships. Key Peruvian NGOs affiliated to Red Puentes Internacional are the Labor Advisory Council of Peru (CEDAL) and the Center of Studies for Development and Participation (CEDEP). CEDAL and CEDEP promote the sustainability of SMEs by incorporating business social responsibility principles in SMEs.

The Business Network is also a global network integrated by platforms and individual NGOs at local, regional and global level. Forum Empresa is the platform of NGOs promoting social responsibility in larger businesses, and their small scale and medium-sized suppliers. Forum Empresa involves NGOs of several Latin American countries, and encourages cooperation, and resource exchange among NGOs, companies and other actors in the network. In Peru the key actor affiliated to Forum Empresa is Peru 2021. Peru 2021, organized in a Board of Trustees, promotes business social responsibility in cooperation with the largest companies in Peru.

6.2.1 The social justice network

Key actors: Labor Advisory Council of Peru

The Labor Advisory Council of Peru (Centro de Asesoría Laboral del Perú, CEDAL in Spanish acronym), founded in 1977, promotes people's human rights in order to contribute to sustainable development, and economic and social justice in Peru. CEDAL implements projects, studies and publications on labor, social, economic and cultural rights, and intends to influence the national government and international human rights bodies in favor of workers (CEDAL website, 2011; Moura, 2008, interview).

CEDAL promotes business social responsibility by encouraging larger business to adopt international standards, such as SA 8000, Global Reporting International (GRI) standards, Ethical Trading Initiative (ETI) standards, and AA1000. CEDAL is the pioneer organization in promoting business social responsibility for small scale enterprises in Peru. Since 2003 CEDAL promotes business social responsibility as strategy to improve social and economic performance of small scale enterprises in order to meet national regulation and international standards on labor rights and good environmental practices. By doing so, CEDAL aims to improve the position of small scale enterprises in the market (Moura, 2008, interview).

CEDAL's involvement with social responsibility has to be traced back to the Consumers International (CI) World Congress in 1997 and the foundation of the Fair Labor Association (FLA) in 1998. However, the starting point of CEDAL in promoting business social responsibility was the elaboration in 1999 in cooperation with the Social Accountability International (SAI), of a set of indicators to measure labor rights in transnational companies operating in Peru (CEDAL website, 2011; Moura, 2008, interview). In 2004 CEDAL elaborated a proposal of certification of good practices in SMEs for the Ministry of Employment of Peru. The proposal did not work out due to the lack of incentives to encourage SMEs to adopt business social responsiblity principles (Moura, 2010, interview). In 2009 CEDAL elaborated an ethical code for NGOs based on the ISO 26000. ISO 26000 is a non-certifying standard of social responsibility, officially released in 2010. The ethical code aimed to account the social performance of NGOs. About 50 NGOs were invited to sign the code, including the NGOs affiliated to Red Puentes Internacional. Only 15 NGOs meet the code. Lately, endorsing the code has become harder, so the initiative was cancelled. The difficulty was the labor issue within NGOs. NGOs were also afraid to be held accountable, since NGOs endorsing the code will periodically be monitored by a committee (Moura, 2008, interview; Moura, 2010, interview). Similarly in Paraguay the construction and implementation of a code for NGOs also failed. In countries like Spain and the Netherlands ethical codes for NGOs are in place. Ethical codes were also intended to be applicable for companies, including SMEs, in Peru. Table 6.1 shows the key local and international partners of CEDAL in promoting business social responsibility for SMEs in Peru.

Key actors: Center of Studies for Development and Participation

The Center of Studies for Development and Participation (Centro de Estudios para el Desarrollo y la Participación, CEDEP in Spanish acronym), founded in 1977, is a NGO specialized in strengthening the social and productive capacities of development agents by elaborating research into rural and urban social-economic development, proposing public policies and encouraging dialogues on human rights, and advancing sustainable development and equity (CEDEP website, 2011).

CEDEP aims to improve productive and managerial capacities of rural and urban small scale enterprises and the sustainable management of renewable natural resources. CEDEP provides financial and organizational capacities to small scale enterprises, including agribusiness, to adopt business social responsibility principles. Such principles include the improvement of working conditions for their employees, the care of workers' health and the implementation of sustainable practices in production in order to increase SME productivity and their access to competitive national and international markets (CEDEP website, 2011; Minanya, 2007, interview; Minaya, 2008, interview). Table 6.2 shows the key local and international agencies that support CEDEP in promoting social responsibility for rural and urban SMEs.

CEDEP has two regional offices: one in Ica, on the southern coast, and one in Ancash, in the central Andes. The national coordination office located in Lima conducts research, coordinates the regional offices and projects, and lobbies the national government. CEDEP implements projects in 7 regions of Peru including Ica, Ancash and Lima (CEDEP website, 2011). Small scale enterprise beneficiaries of CEDEP are placed in three regions of Peru; Lima (districts of Villa El Salvador and Independencia), La Libertad (districts of Trujillo, Porvenir and La Esperanza) and Ica (Districts of Santiago and Pueblo Nuevo). The small workshops producing for the local markets are placed in Lima and Trujillo. The small agri-industries placed in Ica are single SMEs, not producer associations, and their production is mainly oriented to international markets.

Table 6.1. Key local and international partners of CEDAL in promoting business social responsibility for SMEs.

	Organizations/platforms
Peru	Red Puentes Peru
	Peruvian Network of Solidarity Economy (GRESP)
International	Red Puentes Internacional
	OXFAM-INTERMON (Spain)
	AVINA Foundation

Source: Moura, 2008, interview; Moura, 2010, interview.

Table 6.2. Key local and international partners of CEDEP promoting business social responsibility for SMEs.

		Partner organizations
International		OXFAM NOVIB (the Netherlands)
		EED (Germany)
		ACDI –FIG (Canada)
		IDB-MIF (Unites States)
		CEPAL (United Nations agency)
		DIAKONIA (Sweden)
		Embassy of Japan in Peru
		UNDP
National	Private agencies	CIES –IDRC
		PROPUESTA CIUDADANA GROUP
		the Swiss Foundation for Technical Cooperation (SwissContact Peru)
		Antamina Mining Fund
	Govermental agencies	FONDOEMPLEO

Sources: CEDEP website, 2011; Minaya, 2007, interview.

Key actors: Red Puentes Internacional

Red Puentes Internacional, founded in 2002, is an international platform of Latin American, Spanish and Dutch NGOs and other civil society organizations aimed to promote and monitor business social responsibility in Latin American countries. By doing so, Red Puentes Internacional intends to protect human rights, and promotes social justice and sustainable development in the region of Latin America (Moura, 2010, interview; Red Puentes Internacional website, 2011).

Red Puentes affiliate 43 civil society organizations, including NGOs, of Argentina, Brasil, Chile, El Salvador, Spain, Mexico, Peru, The Netherlands and Uruguay (Red Puentes Internacional web site, 2010). The coordination of Red Puentes Internacional is currently based in Argentina (Moura, 2010, interview). National coordinators of Red Puentes Internacional periodically meet to establish the overall international agenda, together with their funding agencies. The May 2010 meeting gathered national coordinators of Red Puentes Internacional to discuss the platform strategic planning for the years 2011-2013 (Declaración RPI, 2010).

Red Puentes Internacional organizes its work in three themes. The first theme monitors good practices of financial institutions, especially banks. This theme is led by the Dutch NGO SOMO. The second theme looks at the relationship between SMEs, business social responsibility and the national governments. NGOs of Peru, Mexico and Uruguay are leading this theme. The last theme deals with monitoring good practices of business through the Information Center of Business Behavior (CICE) (Minaya, 2007, interview; Moura; 2008, interview; Red Puentes Internacional website, 2011).

The Red Puentes Internacional platform has a branch in Peru, called Red Puentes Peru. Red Puentes Peru was founded in 2005 and focuses on issues of environment, labor rights, citizen participation and SMEs. Red Puentes Peru monitors the business social responsibility programs of larger companies, systematizes cases and develops indicators for business social responsibility, builds capacities in SMEs and formulates public policies (Minaya, 2007, interview; Moura; 2008, interview; Red Puentes Internacional website, 2011). Red Puentes Peru works in coordination with Red Puentes Internacional to lobby against the national government, raise public opinion and enhance the capacity of NGO members to influence and negotiate (Minaya, 2007, interview; Minaya, 2008, interview; Moura, 2008, interview). During 2010 the platform has shown scarce activity, mainly due to shortage of funding. OXFAM-NOVIB, the main funding agency providing economic support to the platform, ended its financial cooperation with Red Puentes Internacional in 2010. In 2010 Red Puentes Peru is composed of 9 NGOs. ASPEC is the coordinator. CEDEP was the former coordinator. Table 6.3 shows Red Puentes Peru, including its affiliated NGOs and main international partners.

Network relationships: cooperation

In promoting business social responsibility in SMEs, CEDAL and CEDEP currently collaborate with other NGOs, SMEs, governmental agencies, larger companies, NGO platforms and international cooperation agencies. Regarding cooperation with other local NGOs, CEDAL and CEDEP closely collaborate with COPEME, Red Social and other members of Red Puentes Peru (Minaya, 2008, interview; Moura, 2008, interview).

Table 6.3. Red Puentes Peru (Red Puentes Internacional website, 2011).

Country	Organizations
Red Puentes Peru	Asociación civil labor
	Catholic relief services –Peru (CRS-Peru)
	Center of studies for development and participation (CEDEP)
	Labor advisory council of Peru (CEDAL)
	Institute for social development networks (Red Social)
	Peruvian consortium of private organizations for the promotion of small and medium-sized business development (COPEME)
	Peruvian association of consumers and users (ASPEC)
	Labor program for development (PLADES)
Spain	OXFAM INTERMON
The Netherlands	OXFAM-NOVIB
	ICCO
	SOMO
	Friends of the Earth Netherlands

NGOs affiliated with Red Puentes Peru and Red Puentes International took active participation in the working groups that developed the ISO 26000 in 2010. In total 88 countries participated in the elaboration of the ISO 26000, including Peru. Red Puentes Internacional had the official status of liaison NGO during the standard development. The Committee Peru of the ISO 26000 was represented by CONCYTEC, which had as its members; CEDAL, PLADES, CEDEP, Labor Ilo, GEA, Peru 2021, CGTP, and governmental agencies (Moura, 2010, interview; iso2600peru website, 2010; iso website, 2011). As Moura stated (2010, interview) 'the ISO 26000: 2010 creates a landmark in the standards promoting social and environmental good practices in all kind of organizations'.

Besides the Red Puentes Internacional, CEDAL and CEDEP collaborate with several platforms of NGOs dealing with labor rights and SME development at national and international level. In Peru CEDAL is affiliated, among others, with the Peruvian Network of Fair Trade and Ethical Consumption. At the international level, CEDAL is member of the International Network of the Economic, Social and Cultural Rights (ESCR-Net) (Moura, 2008, interview). In the case of CEDEP, this NGO also collaborates with several national producer organizations, such as the National Convention for Peruvian Agriculture (CONVEAGRO), the Peruvian Network for Action in Alternative Agriculture (RAAA) and the National Association of Organic Producers (ANPE). At local level CEDEP is engaged with several actors, including local authorities, universities, unions and small business associations (CEDEP website, 2011; Minaya, 2008, interview; Moura, 2008, interview).

Regarding cooperation with small scale enterprises, CEDAL and CEDEP help upgrading the social and environmental performance of small scale enterprises in Peru in order to engage them with larger companies (Minaya, 2007, interview; Minaya, 2008, interview; Moura, 2008, interview). The small scale enterprise beneficiaries commercialize their products directly to consumers or supply them to business intermediaries, including domestic medium-sized companies and foreign larger buyers. In 2009 CEDAL implemented business social responsibility principles in about 60 small scale enterprises of garment and handy craft makers, organized in associations and clusters located in several districts of Lima, such as San Juan of Lurigancho and northern Lima, and Gamarra town in the Lima center (Moura, 2008, interview). Small garment enterprise beneficiaries of CEDAL supply to SARA, a Spanish larger garment retailer, and to Nautica, a global garment retailer. The NGO CEDEP cooperates with small and medium-sized agri-industries, small garment workshops, shoemakers, mental workshops and baker shops. In small scale agri-industries, as a strategy to promote business social responsibility, CEDEP implements good agricultural practices (GAP). GAP is a set of 14 principles that are used the European Union to encourage sustainable agriculture (environment, risk management and product quality assurance). Currently GAP has extended as worldwide guideline. In urban small scale enterprises, CEDEP implements the Human Resources and Productivity Management standards, including Total Quality and the Japanese Quality Management 5Ss. In a large pilot project aimed to improve working conditions and productivity in the Peruvian small scale enterprises during 2005-2007, CEDEP reached about 1000 SMEs with basic trainings, and about 100 employees and 48 small agri-industries (Minaya, 2007, interview; Minaya, 2008, interview).

Regarding cooperation with larger business, CEDAL usually monitor companies to check their social and environmental performance. CEDAL has cooperation ties with the garment corporation WAMA SAC, the Coteexport SAC and the corporacion Topitop SAC, which are SMEs producing cloths for national and international markets (Moura, 2010, interview). CEDEP has also cooperated with Elite and Cementos Lima SA, both larger domestic companies, in raising awareness on occupational health in SMEs (Minaya, 2007, interview; Minaya, 2008, interview). CEDAL does not have direct links, neither with local garment retailers (e.g. SAGA Falabella), nor with international garment retailers (e.g. Ripley, SARA and Nautica). However, CEDAL looks at large garment and multinational retailers as potential partners for the future. CEDAL seeks to work together with larger companies to improve social and environmental conditions of small garment business as suppliers (Moura, 2008, interview). In the words of Moura (2010, interview) 'it is known that Ripley buys 98% of cloths from Chinese clothing enterprises and only 2% from Peruvian small scale enterprises. We are searching for cooperation with large clothing company providers in implementing business social responsibility in their small providers in Peru'.

Regarding cooperation with governmental agencies, CEDAL and CEDEP collaborate with public health institutions, universities and municipalities to support SMEs. Municipalities have provided logistical support during pilot projects for SMEs, for instance, in occupational health campaigns (Minaya, 2008, interview; Moura, 2008, interview).

Regarding cooperation with international cooperation agencies, CEDAL and CEDEP have links with OXFAM-INTERMON, OXFAM-NOVIB, ICCO, SOMO, Friends of the Earth, AVINA Foundation, IDB and the Swiss Foundation for Technical Cooperation (SwissContact Peru) (CEDAL website, 2011; CEDEP website, 2011; Moura, 2010, interview; Red Puentes Internacional, 2010). OXFAM-NOVIB and ICCO provide funding, policy guidance and knowledge to the social justice network's regional and national platforms and key actors (ICCO website, 2011; OXFAM-NOVIB website, 2011). SOMO is the coordinator of the platform. The aforementioned agencies provide to CEDAL and CEDEP several resources, such as knowledge and funding to promote and implement business social responsibility in small scale enterprises in Peru. While CEDAL works closely with OXFAM-INTERMON, with other Spanish cooperation agencies (e.g. Local Authority of Madrid-Spain) and with the AVINA Foundation, CEDEP has links with the Inter-American Development Bank (IDB) and SwissContact Peru. With financial support of OXFAM-INTERMON, CEDAL facilitated connections of small garment workshops with medium and large garment companies of San Juan de Lurigancho and the Victoria, in Lima, and set up their committees of social responsibility. This project ended in April 2010. Additionally AVINA Foundation provided funding to CEDAL to develop the aforementioned proposal of the ethical code for NGOs. IDB provided financial support to CEDEP to improve the occupational health of employees and working conditions in SMEs during 2005-2007. IDB financed 80% of the total costs of the project and the small scale enterprise beneficiaries 20%. The project came out as a result of a common interest between CEDEP and IDB in supporting SMEs to increase their competitiveness (Minaya, 2008, interview). CEDEP also has ties with SwissContact Peru that provides technical guidance on occupational health and working conditions in small scale enterprises. International cooperation agencies promoting business social responsibility are connected by ties with economic actors and national governments in their respective countries (ICCO position paper, 2009; OXFAM-NOVIB Annual Report, 2011). In contrast, in Peru CEDAL and CEDEP have scarce connections with economic actors and do not have cooperation with the Peruvian government (Minaya, 2008, interview; Moura, 2008, interview).

During 2010 CEDAL has been preparing projects to get funding for supporting local small scale enterprises. Meanwhile CEDAL has been making consultancies in fairtrade standards for the local garment and textile industry (Moura, 2010, interview). The national general election of 2011, to elect the President of the Republic of Peru, delayed the implementation of projects of CEDAL and CEDEP during 2011.

Network relationships: power

This section analyzes the power relations in the social justice network, focusing on key organizations identified in Peru and their international connections.

CEDEP and CEDAL are the key national nodes of the social justice network. Red Puentes Internacional is the key regional node. OXFAM-NOVIB and the Interchurch Organization for Development Cooperation (ICCO) are the key nodes at global level. CEDEP, CEDAL and the other 44 members of Red Puentes Internacional have equal power in the networks. Members have more or less the same opportunity to raise ideas and to build up the platform agenda. For instance, CEDAL and CEDEP raised the issue of business social responsibility in SMEs as a new theme in the platform (Minanya, 2007, interview; Minaya, 2008, interview). Asymmetrical power relationships are shown between the promoters and the members of Red Puentes Internacional. OXFAM-NOVIB and ICCO hold more power than the other NGOs affiliated to the social justice network. They make the final decision in setting the agenda. However, cooperation between local and international NGOs is present. They work in coordination using the Red Puentes platform to check companies and influence them to adopt business social responsibility standards and guidelines. On the one hand, international NGOs pressure large buyers and large stores, and lobby unions of larger buyers in their home countries to meet emerging social and environmental standards and regulations for European businesses and markets; and they pressure Peruvian SMEs suppliers to adopt those standards. This has been the case of OXFAM-INTERMON in lobbying Spanish garment stores, such as ITINTEC-SARA branding to pressure the garment small supplier Topitop to re-employ 600 fired employees (Moura, 2008, interview). On the other hand, in coordination with OXFAM-INTERMON, OXFAM-NOVIB, SOMO and Friends of the Earth, and with small scale enterprise associations and clusters, single small scale enterprises and unions, Red Puentes Peru organizes campaigns to raise public opinion and lobby the national government on the need of social and environmental responsibility of companies (Moura, 2008, interview).

CEDAL and CEDEP have an asymmetrical relationship with small scale enterprises. While these NGOs provide funding and logistical support to small scale enterprises, small scale enterprises are usually 'passive' beneficiaries. Only in some cases small scale enterprises allocate resources during the implementation of the project. Overall SMEs are not very enthusiastic in receiving those resources. Small scale enterprises, especially the urban enterprises (e.g. Industrial Park of Villa El Salvador), are 'overwhelmed' by NGOs promising support. Increasingly small scale enterprises are not interested in cooperating with NGOs. SMEs claim that NGOs do not deliver what they promise. Often small scale enterprises are engaged in cooperation with NGOs only for the period of project implementation. After it ends, small scale enterprises return to 'business as usual' (Minaya, 2008, interview).

The relationship of NGOs with larger companies is quite different. Larger companies are in better positions to establish the conditions for cooperation than CEDAL and CEDEP. Larger companies have enough resources and several potential partners to implement measures of business social responsibility in supply chains. Larger companies have a wider set of resourceful partners than NGOs, including consultancy companies, business associations and business foundations. In the efforts not to lose ground against larger companies, NGOs established Red Puentes Internacional. Indeed, Red Puentes Internacional emerged as an answer to the setting up of Forum Empresa and FUNDES. FUNDES is an organization of regional scope with national offices in 10 Latin American countries. FUNDES supports SMEs and has developed a methodology for applying business social responsibility for Latin American SMEs. FUNDES works in cooperation with Forum Empresa. Another evidence of the power struggle between NGOs and businesses happened in Peru. CEDAL, CEDEP, COPEME and Red Social established Red Puentes Peru as an answer to the launching of Peru 2021 by Peruvian companies.

6.2.2 The business network

Key actors: Peru 2021

Peru 2021 promotes social responsibility as business strategy for companies operating in Peru. By doing so, Peru 2021 intends to position itself as an agent to reach sustainable development in the country (Peru 2021 website, 2010). Peru 2021 has been founded by business people in 1994. It is composed of a Board of Directors that leads the organization and a Board of Trustees that supports it. The Board of Trustees involves representatives of 52 larger Peruvian companies (Table 6.4). Peru 2021 has a coordination office in Lima and a regional office in Arequipa.

Peru 2021 emerged out of a need of frontrunner Peruvian businesses to implement and promote business social responsibility good practices in Peruvian larger companies and the small scale companies providing them. Peru 2021 develops managerial instruments and provides expertise and guidance to implement business social responsibility as strategy of business management in companies and their value chains. For this purpose, at the core of its work, Peru 2021 promotes ISO 14001, eco-efficiency, sustainability reports, business social responsibility policies and related indicators. Peru 2021 elaborates methodologies tailored for Peruvian companies, implements projects, carries out consultancy, publishes publications, raises awareness among the business community and alliances with business associations on business social responsibility. Peru 2021 also channels financial support, expertise and knowledge to facilitate and stimulate the adoption of business social responsibility by companies in Peru. It does not focus directly on SMEs but on larger companies and their value chains. The main clients of Peru 2021 are larger service companies, industries, transnational branches, NGOs, universities, business associations, international cooperation agencies and local banks (Peru 2021 website, 2010; Rizo-Patron 2006, interview; Rizo-Patron, 2007, interview).

Table 6.4. Companies and organizations linked to Peru 2021 (Peru 2021 website, 2010).

Organizations

Board of Trustees	Alicorp	Kimberly Clark
	Amanco del Perú S.A.	La Viga S.A.
	Amrop Hever	Natura
	Aprenda – Grupo ACP	Nextel del Perú S.A.
	Asociación Atocongo Cementos Lima	Odebrecht
	Backus y Johnston S.A.	Pacific Comunicación Estratégica
	Minera Barrick	Pacífico-Peruano Suiza
	Bayer	Pfizer
	BELCORP	Plastisur S.A.
	Banco de Crédito del Perú – BCP	Prodac Bekaert
	BBVA Banco Continental	ProFuturo AFP
	Banco Interamericano de Finanzas – BIF	Perú Waste Innovation S.A.C.
	Compañía Minera Antamina S.A.	Qualitas del Perú
	Compañía Minera Milpo	Repsol del Perú
	Compañía Minera Poderosa S.A.	Red de Energía del Perú
	Coca-Cola Servicios del Perú S.A.	San Fernando S.A.
	DBM Perú	Securitas del Perú
	Eikon Comunicación e Imagen	Siderpers
	Empresarial S.A.C.	Southern Perú Copper Corporation
	Empresa de Generación Eléctrica de	Telefónica del Perú
	Arequipa – EGASA	Terminar Internacional del Sur
	Empresa Periodística Nacional – EPENSA	Toyota del Perú
	Ernst & Young	Universidad de San Martín de Porres
	Estudio Grau	Votorantim Metais – Cajamarquilla S.A.
	Ferreyros S.A.	Xstrata Tintaya S.A.
	Incalpaca TPX	Minera Yanacocha
	InVita	Yura S.A.
Peru	National Confederation of businesses (CONFIEP)	
	National of Exporters (ADEX)	
	Cámara de Comercio Americana del Perú	
	Cámara de Comercio e Industria de Arequipa	
	Cámara de Comercio de Lima	
	Instituto Peruano de Administración de empresas	
	Pontificia Universidad Católica del Perú	
	Universidad del Pacífico	
	Grupo GEA	

Table 6.4. Continued.

	Organizations
International	Forum Empresa
	Instituto Ethos
	Global Reporting Initiative (GRI)
	Inter-American Development Banck (BID)
	World Business Council for Sustainable Development (WBCSD)
	AVINA Foundation

Key actors: Forum Empresa

Forum Empresa, founded in 1997, is a continental platform of 19 Latin American and North American NGOs, which promotes business social responsibility in the region of Latin America. The platform affiliates about 2,400 companies. Forum Empresa is a platform of NGOs established by business people to facilitate connections and exchange of information and knowledge about business social responsibility in Latin America. Forum Empresa provides funding to their members to implement projects on business social responsibility (Forum Empresa website, 2011). The key members of Forum Empresa are Institute Ethos of Brazil and the Colombian Center of Social Responsibility. The Instituto Ethos is implementing the Latin American Program of Business Social Responsibility (PLARSE). In cooperation with Peru 2021 and other members of Forum Empresa, PLARSE aims to develop indicators to measure business social responsibility in Latin American countries (PLARSE website, 2011). Acción RSE of Chile, one of the members of Forum Empresa, is leading the promotion of business social responsibility in value chains of larger companies, including SMEs (Forum Empresa, 2010). Several intergovernmental organizations, financial institutions, international cooperation agencies, business foundations and business platforms are linked to Forum Empresa (Table 6.5).

Network relationships: cooperation

In this section the relationships in the business network is described, particularly the connections and cooperation of Peru 2021 with other NGOs, business associations, SMEs, universities, national governmental agencies and international cooperation agencies.

Regarding cooperation with NGOs at the Latin American regional level, Peru 2021 cooperates with NGOs affiliated to Forum Empresa. Peru 2021 has strong ties with Brazilian and Chilean NGOs. Together with the Ethos Institute of Brazil, Peru 2021 developed indicators to measure the social performance of companies. Both NGOs have also elaborated case studies on sustainability reports of Peruvian companies. During 2004-2007 Forum Empresa implemented a project to promote social responsibility at larger companies and SMEs in Brazil, Chile, El Salvador and Peru. Peru 2021 received inputs from key NGOs with expertise in the application of business social responsibility principles in SMEs, especially from Chilean NGOs, such as Action RSE, FUNDES

Table 6.5. The Forum Empresa platform (Forum Empresa website, 2011).

	Organizations	
Business NGOs	Fundación del Tucumán (Argentina)	BSR (United States)
	Instituto Argentino de Responsabilidad	FUNDEMÁS (El Salvador)
	Social Empresaria (IARSE)	FUNDAHRSE (Honduras)
	Forética Argentina	Cemefi (México)
	Instituto Ethos (Brasil)	uniRSE (Nicaragua)
	COBORSE (Bolivia)	IntegraRSE (Panamá)
	CBSR (Canada)	ADEC (Paraguay)
	Acción RSE (Chile)	Perú 2021 (Perú)
	CCRE (Colombia)	ConectaRSE (Puerto Rico)
	AED (Costa Rica)	DERES (Uruguay)
	CERES (Ecuador)	Cedice (Venezuela)
Ally	BID	
	ICCO	
	AVINA Foundation	
Sponsors	CEMEX	
	P&G	
	Telefonica	
	HSBC	

Chile, Prohumana and Vincularse. Peru 2021 has also links with the Argentinean Institute of Business Social Responsibility (IARSE) and the Colombian Center of Social Responsibility (Rizo-Patron 2006, interview; Rizo-Patron, 2007, interview; Rizo-Patron, 2008, interview).

Regarding cooperation with NGOs at the national level, Peru 2021 participated together with PLADES and CEDAL in the development of the ISO 26000 standard at the Peruvian working group of the International Standard Organization (Carpio, 2010, interview). Other local NGOs that have cooperative links with Peru 2021 are Gestión ARCE and Responde Peru (Rizo-Patron, 2008, interview).

Several local organizations including NGOs, companies and universities involved in the promotion of business social responsibility in Peru launched a national platform of social responsibility, but it did not stand for long. Peru 2021 and CEDAL were part of the initiative. Referring to this failure, Rizo-Patron (2008, interview) concludes; 'it is hard to work with heterogeneous actors. The platform had a wider agenda and lacked financial support. A similar situation happened with the Global Compact national platform. Peru 2021 prefers to work with platforms that focus on business'.

Regarding cooperation with larger companies, Peru 2021 closely cooperates with larger domestic companies affiliated to the Peru 2021's Board of Trustees. These companies are services companies, banks and larger extractive companies. As part of the aforementioned Forum Empresa's project on business social responsibility, Peru 2021 engaged with 5 larger domestic companies: Backus, Banco de Crédito del Perú, Banco Interamericano de Finanzas, Profuturo AFP and Telefónica del Perú. Peru 2021 supports these 5 companies to manage their business social responsibility programs, particularly elaborating the company's reports on sustainability and indicators of social responsibility. Peru 2021 also works closely with CONFIEP and ADEX, the two main Peruvian business associations (Rizo-Patron, 2007, interview; Rizo-Patron, 2008; interview). Additionally, Peru 2021 has ties with consultancy companies, particularly with Qualitas del Peru and Bureau Veritas (Rizo-Patron, 2008, interview)

Regarding cooperation with SMEs, Peru 2021 only works with SMEs that provide goods and services to larger companies (Peru 2021 website, 2010). During the years 2002-2005 Peru 2021 executed the 'chain' project (the so called 'Cadena Productiva Sostenible Perú 2021', in Spanish), aimed to certify frontrunner SMEs for fulfilling the environmental management system (EMS) standard ISO 14001. SMEs participating in the project had to be linked to a value chain of larger companies. Besides ISO 4001, standards of business social responsibility and eco-efficiency were implemented. Moreover, as part of the aforementioned regional project of Forum Empresa, 41 Latin American SMEs implemented business social responsibility principles. In Peru, as part of the same project, 10 SMEs that provided good and services to larger companies participated: Cadillo Comunicación Gráfica, Corporación de Alta Tecnología – CALTEC, Editorial Pacasmayo, Excellent Courier, EXACT, HGM – Soporte y Soluciones, Innovación Empresarial, Polinomio, TAI HENG and Tecnogas. Peru 2021 collaborated most with SMEs providing services to larger companies, but not with SMEs that are part of productive value chains, such as garment enterprises and agri-industries.

Regarding cooperation with universities, Peru 2021 closely works with the Catolica University's Institute of Quality and the Pacifico University's Research Center (CIUP) in public awareness and elaboration of studies on business social responsibility. In cooperation with both universities, Peru 2021 organizes fairs on annual bases (the so called 'Expoferia') on business social responsibility in the cities of Lima, Trujillo and Arequipa (Rizo-Patron, 2006, interview).

With respect to cooperation with national governmental agencies, Peru 2021 has ties with the Ministry of Environment (MINAM) and the Ministry of Work and Employment Promotion's Center for the Promotion of Small and Micro Enterprises (PROMPYME) (Rizo-Patron, 2006, interview; Rizo-Patron, 2007, interview; Rizo-Patron, 2008; interview).

Peru 2021 is linked with several international cooperation agencies: the inter-American development bank (IDB), the Multilateral Investment Fund (MIF), the Organization of American States (OAS), the Canadian International Development Agency (CIDA), AVINA Foundation, ICCO, the Global Compact Initiative, the Business for Social Responsibility platform (BSR), Price Waterhouse Coopers, Swiss contact Peru and the World Business Council for Sustainable Development (WBCSD) (Peru 2021 website, 2010; Rizo-Patron, 2006, interview; Rizo-Patron, 2007, interview; Rizo-Patron, 2008; interview). IDB has provided support to Peru 2021 and to Forum Empresa in terms of policy guidance and funding to promote sustainable business at the Latin American regional level, particularly for SMEs. The IDB policies on business development at regional level are operationalized through MIF (IDB website, 2011). IDB was the pioneer agency promoting engagement of SMEs suppliers in business social responsibility efforts of larger buyers (IDB website, 2011). It also provided financial support to Peru 2021 for the 'chain' project (2002-2005) and for the Forum Empresa regional project (2004-2007). In 2007 Peru 2021 applied to IDB for funding to implement social responsibility in value chains in Peruvian larger companies, including their SME providers but the application did not work out (Rizo-Patron, 2007, interview; Rizo-Patron, 2008, interview). Peru 2021 applied again for IDB funding in 2010, together with other members of Forum Empresa, to evaluate the SMEs participating in the former regional project (Carpio, 2010, interview). The national presidential election of 2011 has delayed the start of that project.

In 2001 Peru 2021 became the national node of the global platform WBCSD. WBCSD is the international platform of companies to explore sustainable development, share knowledge and best practices, and to advocate the business position in a variety of forums in the five continents (Peru 2021 website, 2010; Rizo-Patron, 2006, interview; Rizo-Patron, 2007, interview; Rizo-Patron, 2008; interview).

Network relationships: power

This section analyzes the power relations in the business network, focusing on the key organizations identified in Peru and their international connections. Peru 2021 is the key national node of this business network. The regional nodes are the Inter-American Development Bank (IDB), the Organization of American States (OAS) and Forum Empresa. The key global node is the World Business Council for Sustainable Development (WBCSD).

Peru 2021 has symmetrical power relations with larger domestic companies in terms of exchange of resources. For instance, establishing the institutional agenda of Peru 2021 is done in coordination with the Board of Trustees' companies (Rizo-Patron, 2008, interview). Moreover, Peru 2021 works in alliance with the Board of Trustees' larger companies to influence the Peruvian business community towards sustainable practices (Peru 2021 website, 2010). Peru 2021 and the Board of Trustees' companies work on a mutual benefit basis. While Peru 2021 provides expertise and funding for larger companies to implement business social responsible projects, larger companies need Peru 2021 to help them to implement and advertise their business social responsibility initiatives and to safeguard their 'licence to operate'. SMEs have asymmetrical power relations with their buyers and with Peru 2021. Larger companies govern the value chain and set the conditions for their SME providers, as shown in the business social responsibility projects involving SME suppliers. Small scale enterprises are passive recipients of resources and conditions in the business network.

Peru 2021 and the other members of Forum Empresa have more symmetrical power relations. Forum Empresa not only facilitates the exchange of information and knowledge among its members but also connects its members with international cooperation agencies, for instance, the aforementioned transfer of expertise from Chilean NGOs to Peru 2021 on the application of business social responsibility for SMEs. Additionally, the fact that in the business network business representatives are part of NGOs and NGOs closely work with companies minimizes conflicts.

International cooperation agencies, project operators and SME beneficiaries have asymmetrical power relations in the business network. Regional and national key actors perform as project operators. They channel resources from global nodes to larger domestic companies, including SME providers. The power position in the business network of, for instance, IDB and WBCSD, is based on their capacity to provide resources in terms of finances, knowledge, agenda setting, policy guidance, managerial instruments and credibility, to affiliated and allied NGOs and network actors operating at the national level. Peru 2021 operationalizes policies from IDB and WBCSD to support larger domestic companies and SMEs to adopt business social responsibility principles (Peru 2021 website, 2010). Asymmetrical power relations among actors in the business network are accepted by actors as a given division of tasks with mutual benefits.

6.2.3 Networks: patterns, challenges and trends

In this section the main patterns, challenges and trends of the social justice network and the business network are described, under the heading of cooperation, power relations, challenges and trends.

Cooperation

Cooperation of actors in the networks of business social responsibility has three characteristics: (1) cooperation is limited to the network members; (2) cooperation depends on the density of connections; and (3) cooperation is intermediated by local NGOs.

Cooperation of actors is limited to their own networks, either the social justice network or the business network. In the case of Peru 2021 cooperation is restricted to NGOs working closely with companies. CEDAL and CEDEP work together as they are part of the social justice network, but do not have ties with Peru 2021 that belong to the business network. CEDAL and CEDEP prefer to work with other NGOs rather than emphasize checking the social and environmental performance of larger companies. Red Puentes Internacional and Forum Empresa are parallel platforms in Latin America. Similarly, Red Puentes Peru and the Peru 2021's Board of Trustees are parallel platforms promoting business social responsibility in Peru. It is worth mentioning that lack of cooperation between both networks is especially relevant at the institutional level. At the individual level both networks have collaborative ties. For instance, NGOs staff professionals of both networks usually exchange information. And NGOs of Red Puentes and Peru 2021 worked together to elaborate the Peruvian proposal for developing the ISO 26000. Furthermore, each network involves different types of NGOs. The social justice network involves NGOs with a long history, working usually on social justice issues. The involvement of social justice NGOs with SMEs and supply chains of larger companies originate from the last decade. The business network involves NGOs closely working with larger companies. NGOs in both networks are expanding their connections beyond their traditional social sphere. In the case of the social justice network, NGOs used to be engaged only with human rights and social movement organizations but they are now expanding their connections towards economic actors. The targets are individual small scale enterprises, and clusters and associations of small scale enterprises producing for local and international markets. An example is the motivation of CEDAL to explore links with transnational buyers in the clothing industry. A summary of the main actors and platforms of the networks of business social responsibility is presented in Table 6.6. An overview of the business social responsibility networks is presented in Figure 6.1.

Table 6.6. Main actors and platforms in business social responsibility networks for SMEs in Peru.

Networks	National actors	National platforms	Regional platforms	Regional/global actors	Global platform
Social Justice	CEDAL, CEDEP	Red Puentes Peru	Red Puentes Internacional	OXFAM-NOVIB, ICCO	–
Business	Peru 2021	Board of Trustees	Forum Empresa	IDB	WBCSD

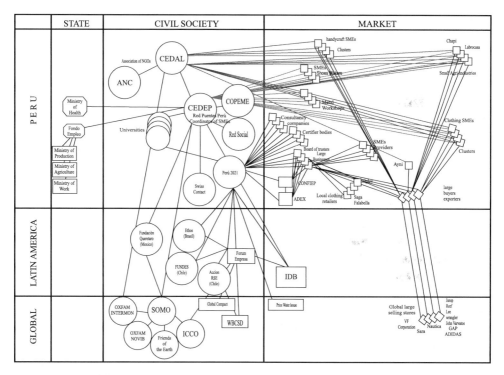

Figure 6.1. Map of the networks of business social responsibility, including key actors, platforms and connections.

The second characteristic of these networks is that they are densely connected at the global pole and scarcely connected at local pole. IDB, ICCO and OXFAM-NOVIB, among others, exchange resources more intensely than local NGOs and SMEs located at local pole. Furthermore, common connections to both networks are also found at the global pole. For instance, IBD and ICCO are part of both networks. IDB has funded projects of CEDEP, Forum Empresa and Peru 2021. ICCO also provides funding to Red Puentes Internacional and has connections with IDB. Furthermore, most SMEs, both rural and urban enterprises, including those with certain economic stability, are isolated from each other with scarce ties among them. SME associations are disconnected and segmented, especially in those of urban small scale enterprises, making connections with the networks of social responsibility harder. SMEs in the social justice network do look for links with larger companies. For instance, clusters and associations of clothing and handcraft small scale enterprises are recently emerging in the local market looking for ties with larger buyers. Finally, at the local pole the social justice network actors (e.g. CEDEP and OXFAM/NOVIB) have connections with actors (e.g. ANPE, RAAA) of the organic production networks. However, those connections do not end up in collaborative projects yet.

Finally, cooperation between SMEs and larger companies and between SMEs and international cooperation agencies is intermediated by local NGOs. For instance, CEDAL and CEDEP cooperate with SMEs to connect them with local larger and medium-sized buyers and exporters for international markets in the garment, handicraft and food industry. Peru 2021 works in close cooperation with larger companies incorporating standards of business social responsibility at the management level of the company and their SME suppliers. International cooperation agencies are not able to work directly with SMEs, as they need the intermediation of local NGOs to reach SMEs.

Power relations

Power relations of actors in the networks of business social responsiblity have three characteristics: (1) they are asymmetrical power relations; (2) local NGOs strongly depend on global nodes; and (3) key network actors and SMEs have different interests.

International cooperation agencies, local NGOs and SMEs have asymmetrical relations with unequal distribution of power. At one pole of the networks international cooperation agencies are located, with capacities to set the network agenda, guide policies and provide funding. The other pole locates small scale enterprises with scarce ties among each other and with market actors. Although local NGOs are the mediators between both poles, they have scarce power in the networks and hardly establish agendas and influence policy guidelines. In the case of the platforms, Forum Empresa is less asymmetrical than Red Puentes Internacional regarding the relations among their affiliated organizations. Peru 2021 has more equal power relations with larger domestic companies. Finally, international cooperation agencies and local NGOs show strong top-down power relations towards SMEs in both networks.

Secondly, local NGOs are self-interested organizations competing for funding resources, especially from international cooperation agencies to implement projects of business social responsibility in SMEs. NGOs affiliated with the social justice network present higher degree of dependency on funding than NGOs affiliated with the business network. NGOs of the business network have more diversified sources of funding, as they also receive funding from larger companies, business associations, business foundations and international NGOs.

Third, neither are SMEs interested in gaining power positions in the networks, nor are the networks interested in including SMEs as actors. This mismatch is caused by different interests of SMEs and key network actors. While most SMEs perform in the frame of 'business as usual' aiming to get enough income to survive, key actors of both networks are interested in improving the social and environmental downsides of global value chains. Only SMEs that are part of exporting supply chains or SMEs that are eager to become suppliers are included in the networks as beneficiaries. The social justice network and the business network do not have links with the majority of SMEs in

Peru, as NGOs only reach SMEs that are suppliers, or have enough capabilities to become suppliers, of larger companies. International cooperation agencies, in cooperation with international and local NGOs design policies, offer funding and allocate expertise, but many SMEs are reluctant to collaborate and prefer to keep at distance from the networks of social responsibility. Most SMEs scarcely buy-in business social responsibility, though promoters make tremendous effort in promoting it. SMEs take the position of passive beneficiaries. As a way to gain legitimacy with their SME beneficiaries, NGOs are trying to act by example with only limited success. For instance, CEDAL and other like-minded NGOs have encouraged the NGO community of Peru to adopt an ethical code of social accountability, but most NGOs did not adopt it. Moreover, micro and small scale enterprises supplying to local markets are excluded from the agenda of both networks. This means that the networks of business social responsibility exclude most SMEs, missing the opportunity for furthering social responsibility in the entire local business community in Peru.

6.2.4 Conclusions

a. The social justice network and the business network are the main networks involved in the promotion of business social responsibility for SMEs in Peru. Despite the fact that both networks are working on the same issue, they do not have ties and thus only limited cooperation. This makes cooperation very selective and limited to the network members, especially at the local and regional level. Networks promoting business social responsibility are structured in platforms at national, regional and global level. At each level there are key actors. At the national level key actors are local NGOs and larger companies. SME beneficiaries include small scale suppliers of larger companies, associations and clusters of SMEs in sectors of interest for international markets. Platforms at the national level are Red Puentes Peru and the Peru 2021's Board of Trustees. At the regional level the platforms are Red Puentes International and Forum Empresa. International cooperation agencies are the main promoters of business social responsibility at regional and global levels.

b. Power relations between network nodes and SMEs are asymmetrical. Local NGOs strongly depend on global nodes for resources to perform their intermediary roles in the networks. Larger companies have more balanced power relations with business NGOs than with social justice NGOs. Neither are SMEs interested in gaining power positions in the networks, nor do the networks offer conditions to include them as legitimate actors in the networks. The network global nodes are only interested in building up cooperation with small scale enterprises that are capable to be suppliers of larger buyers. There is no interest in micro and small scale enterprises oriented to local markets. Small scale suppliers rely on international cooperation agencies and larger buyers to improve their social and environmental downsides, but they are just passive beneficiaries.

c. Two types of NGOs are identified in the networks: social justice NGOs and business NGOs. While social justice NGOs are in a better position to be linked with SMEs, business NGOs are in better position to be linked with larger companies. Social justice NGOs are more involved with clusters of small scale garment, agri-industries and shoemaker workshops eager to be part of global supply chains; or with larger buyers interested in incorporating sustainable practices in their small scale suppliers. Business NGOs usually collaborate with insurance, financing and natural resources extractive companies and their supply chains. However, both types of NGOs look for possibilities to influence larger companies and their small scale providers to adopt business social responsibility. This finding provides evidence that actors of both networks are expanding their ties. While social justice NGOs are moving from social to economic roles, companies and social justice NGOs are moving from economic to social roles. At the meantime, small scale enterprises are getting closer to larger companies by becoming their suppliers.

d. The most important challenge that NGOs are facing in promoting business social responsibility for SMEs, next to the limitations of expertise and funding, is to convince SMEs to adopt business social responsibility standards and guidelines. Most SMEs lack of motivation to adopt them. SMEs are also reluctant to collaborate with NGOs due to the mistrust towards them. This is so because SMEs are excluded from decision making by the key network actors and platforms. Finally, NGOs have limited connection with the social nets of SMEs. Increasing ties with SMEs might likely contribute to cooperation between the social justice network and business network.

6.3 Discourse analysis of business social responsibility

6.3.1 Introduction

NGOs are very active agents in promoting business social responsibility for SMEs. Therefore, analyzing the perspectives of agents involved in such promotion becomes relevant. NGOs follow a particular set of views to reach SMEs. Those views are clustered in two discourses: the discourse of 'business upgrading' and the discourse of 'corporate responsibility'. Additionally, NGOs endorsing the business social responsibility discourses can be clustered in three groups: 'Traditional' NGOs, 'technocratic' NGOs and 'business' NGOs.

Based on the Dryzek's (1997) definition of discourse, business social responsibility can be seen as a diverse set of assumptions, judgments and contentions. Each discourse of business social responsibility is a shared way to apprehending improvement on production efficiency encompassing plausible storylines or accounts about production and the market by agents including NGOs, SMEs and micro-enterprises, business associations, governmental agencies and international cooperation agencies. The discourse analysis of business social responsibility focuses on views about production, the market and environmental sustainability, and the role of key agents involved in making production of SMEs sustainable.

The discourses of business social responsibility are analyzed following the methodology of Dryzek (1997), as described in depth in Chapter 3. Four elements are analyzed to construct storylines: (1) basic entities acknowledged; (2) assumptions about natural relationships; (3) agents and their motives; and (4) key metaphors and other rhetorical devices.

Discourses of business social responsibility

The discourse of 'business upgrading' and the discourse of 'corporate responsibility' have communalities and differences in their views (Table 6.7). The communalities are that both discourses identify the same key actors: larger companies, international cooperation agencies and NGOs.

Also, both discourses consider the role of larger companies central in triggering improvement of small suppliers in value chains. The difference relates to the focus of each discourse. While the business upgrading discourse focuses on small scale enterprises, the corporate responsibility discourse focuses on large companies.

The business upgrading discourse encompasses the idea of business social responsibility as a strategy that matches the perspectives of economic and social priorities with that of sustainability. The discourse focuses on small scale enterprises engaged in low and high profitable value chains. The main motivations of agents are to build connections of small scale enterprises with sustainable value chains and to influence 'conventional' value chains to become sustainable. 'Traditional' NGOs and small scale enterprises are forced to change their perspectives about traditional ways to conceptualize business. Business upgrading highlights the idea of cooperation between NGOs and companies to support small scale providers. Internalizing social and environmental concerns in small scale enterprises is seen as key contribution to their competitiveness.

The corporate responsibility discourse encompasses the idea of business social responsibility as a strategy of larger companies to become sustainable. Larger companies are seen as 'pullers' of small scale providers of profitable value chains. Most small scale enterprises that are part of informal and low profitable value chains are seen as incapable to adopt business social responsibility. Only small scale providers attached to profitable value chains have capacities to adopt social and environmental

Table 6.7. Overview of business social responsibility discourses.

Discourse	Business upgrading	Corporate responsibility
Storylines	Business social responsibility is seen as the strategy to match economic and social rights with sustainability. Connecting small scale enterprises with larger companies and influencing them to become sustainable are central in the discourse	Business social responsibility is seen as strategy that contributes to the sustainability of larger companies and their supply value chains. Only small providers of profitable value chains have capacities to adopt social and environmental standards
Type of NGOs	Conventional NGOs	Business NGOs, technocratic NGOs
Subscriber	CEDAL, CEDEP, APOMIPE	Peru 2021, Responde, Impulsa Peru, APCPROCALVIDA, CAJ, DESCO, GEA-Perú, SASE Asociación civil, CER

standards and guidelines. The discourse focuses on profitable larger companies and their small scale providers. The main motivation of agents is to make larger companies sustainable at the long-term. NGOs cooperate with larger companies and share perspectives about the social role of business. 'Technocratic' NGOs and 'business' NGOs subscribe to this discourse.

The business upgrading discourse and the corporate responsibility discourse came out of radical opposing discourses. The business upgrading discourse has its roots in the 'rights' discourse. The corporate responsibility discourse has its origin in the 'liberal market' discourse. The rights discourse embodies the idea that powerful actors, such as larger companies have self-interested motivations and potentially are able to harm society by abusing basic human rights of citizens and workers (ESCR-Net website, 2010; Lusiani & Feeney, 2009). Rights involve not only the political rights of freedom to speech, but also labor and cultural rights (OHCHR website, 2010; UN website, 2010). In this discourse, companies mistrust NGOs and unions. Unions also mistrust companies. So, companies, unions and NGOs are involved in an 'adversarial' culture. The adversarial culture involves struggling, conflicting and opposing views between companies as 'the exploiters' and workers as the 'exploited'. As NGOs have a tradition of defending labor rights of workers, companies perceive them as adversaries. In the words of Moura (2008, interview) 'unions understand business social responsibility as the responsibility of the employers to improve the salary of workers. As long as workers obtain higher salary they do not care about the environment. Unions are only focusing on labor rights and the International Labor Organization (ILO) regulations in the struggle against their transnational employers'. 'Pressure' and 'demands' are key rhetorical devices in this discourse. In contrast, the liberal market discourse embodies the idea that the owning and operating of the means of production by businesses form the bases for societal welfare. In a statement, the WTO's Director-General (WTO website, 2010) stressed that 'the opening of national markets to international trade, with justifiable exceptions or with adequate flexibilities, will encourage and contribute to sustainable development, raise people's welfare, reduce poverty, and foster peace and stability'. The liberal market discourse considers the market driven supply and demand central mechanisms to determine investments, distribution, income and pricing of goods and services. Companies are central actors and are the engines that deliver goods and services for satisfying the demands of consumers. As individuals seek for their own benefit in society by their own means, the role of institutions, particularly the national government and NGOs, are to set the stage for market actors. 'Freedom' and 'growth' are key rhetorical devices in this discourse.

In the next sections, the discourse of business upgrading and the discourse of corporate responsibility are developed in more detail.

6.3.2 Business upgrading discourse

Let's improve small business competitiveness by upgrading their social and environmental performance!

Business upgrading is seen as an opportunity for improving the competitiveness of small scale enterprises. The upgrading is in terms of adopting social and environmental standards by small scale enterprises to be more competitive in the market. 'Brave' small scale enterprises, individual ones or organized in clusters or associations, are able to gain access to new markets and connect with social responsible value chains. An overview of the discourse of business upgrading is summarized in Table 6.8.

Basic entities acknowledged

The liberal market is acknowledged. The liberal market is understood as consisting of global chains of transnational companies and financial institutions with large amount of resources and worldwide connections. It encompasses competitive local and international markets connected in global value chains. As the option to operate outside the liberal market boundaries becomes increasingly less feasible, small scale enterprises have no other choice than to move for upgrading their business and connect to transnational company's value chains. The upgrading of business is seen as a compulsory condition for small scale enterprises to access competitive markets, since it implies sophisticated rules and requirements.

Business social responsibility is also acknowledged. Business social responsibility is understood as a strategy to upgrade social, environmental and economic performance of small scale enterprises. Upgrading includes, for instance, increase the standardization of production, eco-efficiency,

Table 6.8. Business upgrading discourse.

Elements	Business upgrading
Basic entities recognized or constructed	Liberal market
	Business social responsibility
	Social and environmental standards, guidelines and ethical codes
	International cooperation agencies, the Red Puentes, transnational companies, municipalities and the national government
Assumptions about natural in relationships	Solidarity
	Hierarchy
	Nature is understood in economic terms and environment as synonym of working conditions
Agents and their motives	International cooperation agencies: competitiveness and sustainability of small scale enterprises
	NGOs, both international and local: support small scale enterprises to meet social and environmental requirements of competitive markets
	Larger companies: keep their position at the market
	Frontrunner small scale enterprises: market expansion
	National government: irresponsible agent
Key metaphors and other rhetorical devices	Profitable markets are something 'vital' for the surviving of small scale enterprises
	'Human rights', 'vulnerable social groups' and 'global standards', 'codes of conduct', 'voluntary norms', 'ethical norms' and 'self-regulation management standards'

improvement of working conditions, occupational health and safety and risk prevention, and labor and environmental standards adoption. The upgrading enhances the possibility of small scale enterprises to access new markets and become suppliers of profitable value chains. The discourse highlights the idea that without considering access to profitable markets, it is impossible for small scale enterprises to grow. In the discourse, growth means not only incremental accumulation of capital and economic stability in the market, but also social and environmental upgrading. Competitiveness and sustainability of small scale enterprises are seen as requirement for long-term profit making. The improvement of the quality of employment, the working conditions and the skills of workers is seen as central in this discourse. In the words of Minaya (2008, interview) 'improvement of economic factors, such as market information, financing and managerial instruments contribute to economic growth of SMEs. But those factors do not necessarily impact in their social performance, for instance, in the quality of employment that small scale enterprises offer'. In the words of Moura (2008, interview), 'business social responsibility helps to widen the usually narrow way to approach the growth of SMEs and to understand the development of small scale enterprise in terms of their integral sustainability at the long-term'. The business upgrading discourse is seen in line with the liberal market discourse in terms of considering companies central in economy but highlighting their social and environmental performance.

Social and environmental standards, guidelines and ethical codes are acknowledged as voluntary instruments to measure the social and environmental performance, including among others SA 8000, ISO 14001, OSHA, GRI, Dow Jones Sustainability Indexes, Global Compact and Millennium Development Goals. The adoption of standards, guidelines and ethical codes are the only feasible way to legitimatize companies within the liberal market economy. As Moura stated (2008, interview) 'the standardization of production in small scale enterprises is a necessary requirement to survive global market pressure'. Standards are seen as departure point for upgrading small scale enterprises and cooperation with stakeholders. The adoption of standards and social responsibility good practices is not seen as a privilege of larger companies, but as a reachable aim for small scale enterprises. Social responsibility in terms of protecting and promoting human rights is applicable to all businesses regardless their size and type. According to the discourse proponents, implementing socially responsible principles does not imply necessarily high investments. Several small scale enterprises in Peru have been identified with good social responsible practices, for instance Ice cream Holanda, the clothing enterprise El Ayni, the agri-industries Chapi and Labrocasa. Another example is the emergence of ethical banks, such as Popolo bank and Triods bank that finance socially responsible small businesses (Moura, 2008, interview). Also, the launching of the ISO 26000 in 2010 is considered a landmark in the promotion of social responsibility worldwide, because it becomes applicable for all types of organizations, beyond businesses.

Other basic entities recognized in this discourse are international cooperation agencies, the Red Puentes platform, transnational companies, municipalities and the national government.

Assumptions about natural relations

Solidarity is at the core of the business upgrading discourse. Subscriber organizations are moved by solidarity values towards small scale enterprises. NGOs see themselves as protectors of social and economic rights of vulnerable people. This role positions NGOs as intermediaries between international cooperation agencies and small scale enterprises. Solidarity is not seen as charity but as support to economic development of disadvantaged small scale enterprises. Business social responsibility is seen as a common framework that facilitates cooperation between companies, NGOs, unions and other actors to improve economic, social and environmental performance of small scale enterprises and engage them in global supply chains.

Relations between organizations are hierarchical based on connectedness. Organizations such as larger companies with dense connections with global actors and NGOs are considered powerful actors and are located at the top. Organizations such as small scale enterprises with scarce connections with global actors are considered less powerful actors located at the bottom.

The relationship with nature is understood instrumentally: nature is input for production. The term 'environment' is frequently used to refer to the good conditions for business competitiveness such as safety, occupational health and risk prevention. Environmental protection is seen as a factor that contributes to improve performance of human resources within small scale enterprises. The ultimate aim of environmental protection is business competitiveness and economic sustainability of small scale enterprises.

Agents and their motives

Agency is granted to collective actors working in networks of global reach. International cooperation agencies and NGOs have agency. Agents elaborate policies, strategies and projects in mutual cooperation to trigger the upgrading and increase of competitiveness and sustainability of small scale enterprises. NGOs endorsing this discourse match the international agreements on human rights and the standards on business social responsibility. In this way, based on social and globally accepted environmental standards, NGOs aim to collaborate with economic actors, break down the adversarial culture against business and go beyond governmental regulation. NGOs subscribing this discourse, on the one hand, aim to build clusters and associations of frontrunning small scale enterprises and connections with larger buyers. On the other hand, NGOs aim to support larger companies and their small providers to adopt socially responsible codes of conduct and standards in the whole value chain (Minaya, 2008, interview; Moura, 2008, interview). Traditional strategies of supporting SMEs, such as formalization, micro-credits and product quality improvement, are seen as attached to socially responsible management of small scale enterprises. Moreover, NGOs are involved in monitoring the claims of transnational companies and their local subsidiaries regarding their commitment on social responsibility and compliance to national regulation. This means that NGOs seek to influence larger companies and their suppliers by monitoring their social and environmental performance. Finally, NGOs build capacities and raise awareness about social responsibility and its advantages for small scale enterprises and small providers to and unions of larger companies.

Larger companies are considered allies of NGOs in the improvement of social and environmental conditions of SMEs. Socially responsible larger companies seek to improve their position in the global market, meet international standards, access niche markets, and meet the demands of pressure groups, watch dog organizations and consumers. Socially responsible companies have higher value at markets, higher rewards from consumers and gain social legitimacy. Retailers, buyers and dealers are central in the discourse. Companies that are not socially responsible risk their brand prestige, for instance when a small provider of a transnational company violates the ban on child labor (Moura, 2008, interview). Companies become socially responsible due to pressure from stakeholders.

Most small scale enterprises are not powerful agents. Small scale enterprises with agency capacities are organized in clusters and associations to supply value change of larger companies. This is the case with small scale garment enterprises, small scale handicraft enterprises and shoemaker enterprises (Minaya, 2008, interview; Moura, 2008, interview). Small and medium-sized agri-industries subscribing to the business upgrading discourse are individual companies (Minaya, 2008, interview). Small scale enterprises that are part of this discourse are in the process of business growth and eager to expand their markets. For instance, APOMIPE aims to increase the competitiveness of small scale enterprises by clustering and connecting them to value chains of larger buyers (APOMIPE, 2007). Although APOMIPE focuses on the business upgrading of SMEs, its intervention is biased to market views rather than to the rights approach. The motivation of small scale enterprises with agency capacities is economic growth, profit making and meeting requirements of their buyers.

National government does not take its responsibility in leading the implementation of national policies for more socially responsible business in Peru. According to the discourse proponents, there is lack of public policies that stimulate business social responsibility, promote cooperation among small scale enterprises, connect small scale enterprises with sustainable value chains and support small scale enterprises that have good practices. In a regional study that includes Peru, Cici and Ranghieri (2008) conclude that environmental damage is considered a serious problem, but not a priority for the national government. Discourse proponents claim that the national government is eager to gather taxes from small scale enterprises, but not no invest in improving their social and environmental conditions. According to Moura (2008, interview), the Peruvian tax agency, SUNAT, is seen as a 'threat' by Peruvian small scale enterprises.

Key metaphors and their rhetorical devices

The business upgrading discourse includes a metaphor that intends to capture the central importance of profitable markets for small scale enterprises. Profitable markets are characterized as something 'vital' for the surviving of small scale enterprises. This notion implies that small scale enterprises cannot grow if they are out of global markets. So, subscribers see business social responsibility as a good strategy for business upgrading of small scale enterprises and connecting them to international buyers and retailers.

The key rhetorical devices in the discourse of business upgrading are 'human rights', 'vulnerable social groups' and 'global standards'. The protection of human rights of vulnerable social groups, particularly the social and economic rights, is at the core of NGOs subscribing this discourse. Global standards refer to social and environmental standards that companies subscribe to, to show their social commitment. NGOs either use the global standards to support the business upgrading of small scale enterprises or to monitor value chains of larger companies. Global standards are related to other rhetorical terms including 'codes of conduct', 'voluntary norms', 'ethical norms' and 'self-regulation management standards'.

6.3.3 Corporate responsibility discourse

Larger companies are pulling small scale enterprises!

Business social responsibility is seen as an opportunity for improving competitiveness of larger companies. The adoption of social and environmental standards and ethical codes impact on the total performance of business. Only larger companies have the capacities to adopt such sophisticated standards and, by doing so, pull their small providers. An overview of the analysis of the discourse of corporate responsibility given in Table 6.9.

Basic entities acknowledged

The liberal market is acknowledged. The liberal market sets the conditions for market access. Access to profitable markets means to be economically competitive and meet the established requirements of economic actors. According to the discourse proponents, the liberal market brings favorable conditions for company competitiveness. Free Trade Agreements in Latin America are seen as a good opportunity for domestic companies to grow. In contrast, local markets are seen as an inefficient place for business growth. Local markets are also seen as disconnected from

Table 6.9. Corporate responsibility discourse.

Elements	Corporate responsibility
Basic entities recognized or constructed	Liberal market Business social responsibility Social and environmental standards, guidelines and ethical conduct codes Sustainability
Assumptions about natural in relationships	Cooperation Hierarchy Nature is seen as input for production Environment is considered as business stakeholder
Agents and their motives	Business NGOs: support larger companies in adopting business social responsibility standards Large companies: pressure, market, image Small providers: economic growth, competitiveness and access to new markets Most small scale enterprises do not have agency at all: incapable to implement business social responsibility standards National government is an ineffective agent
Key metaphors and other rhetorical devices	'Competitiveness', 'stakeholders', 'small providers', 'codes of conduct', 'social balance', 'sustainability' and 'social and environmental standards'

global competitive markets. According to discourse proponents, the liberal market is able to encourage business for change and innovation and has the capacity to connect larger companies with competitive small scale enterprises, and the competitive small scale enterprises with surviving micro scale enterprises. According to De la Torre (2004) 'a liberal market system encourages innovation and wealth creation'. Subscribers to this discourse have shared views regarding the way economic production and trading should be handled.

Business social responsibility is realized in this discourse. As De la Torre stated (2004) 'business social responsibility has the power to develop further stability, harmony and equity in the social structure'. According De la Torre (2004) the allocation of technology and means of production will help to incorporate once excluded Peruvians in modernity. So, the Peruvian society might benefit from globalization. According to the discourse proponents, business social responsibility includes ethical, economic, social and environmental considerations that have to be part of the strategic management of the company (Rizo-Patron, 2007, interview). Business social responsibility helps competitive companies to adapt to the new requirements of global markets and to further competitiveness (Le Bienvenu, 2007). Moreover, business social responsibility of larger companies implies management of expectation of stakeholders (Rizo-Patron, 2006, interview). Business social responsibility is seen as not applicable for most small scale enterprises since they first need to meet basic business management capacities and comply with governmental regulation.

Social and environmental standards, guidelines and ethical codes are acknowledged, such as SA 8000, ISO 14001, OSHA, GRI, Dow Jones Sustainability Indexes, Global Compact and Millennium Development Goals (Canessa & Garcia, 2005). Business social responsibility standards and ethical codes are audited by third parties, not by NGOs. Moreover, the discourse of corporate responsibility includes cleaner production and other tools that contribute to improve environmental performance of SMEs (Schwalb & Malca, 2004; Alegre, 2010, interview).

Sustainability is another entity that is acknowledged in this discourse. Sustainability views are founded in the Brundtland report of sustainable development (Rizo-Patron, 2007, interview). Business social responsibility is a small part of the sustainable development paradigm, focusing on business actors. Economic sustainability at the long-term is at the core of this discourse. Therefore, sustainability is understood as the continue availability of natural resources and the continue buying of products by consumers in such a way that the cycle of production and consumption keeps functioning forever. Economic sustainability is usually seen in terms of economic growth. According to Forum Empresa (2009, interview), 'the effects of incorporating sustainability into value chains not only benefit larger companies and their suppliers, but also more generally reaches people. Due to the relevance of SME businesses in terms of employment, these effects typically have a direct impact on people's lives'. Environmental sustainability is seen as institutional capacity to handle climate change, natural disasters, renewable energy and lower carbon emissions (Forum Empresa, 2009).

Assumptions about natural relations

Cooperation is restricted to actors subscribing to the general viewpoints of the corporate responsibility discourse. Cooperation is mainly done among business NGOs, larger companies, consultancies and other actors subscribing to the discourse. Paternalistic approaches of cooperation are criticized by the subscribers of the discourse. Instead, the adoption of social responsibility is seen as in the interest of larger companies and small providers to improve competitiveness at the market. According Le Bienvenu (2007), failing companies in terms of profit making cannot adopt social responsibility standards. In the past, he claims, companies used to give philanthropic donations. Now, the central and only aim of companies is the creation of value.

Hierarchy is based on formal expertise and economic power of companies. At the top are transnational companies and financial institutions with expertise and capital. At the bottom are small providers engaged in value chains. The intermediaries between both poles are business NGOs, helping larger companies and their small providers to adopt social responsibility standards.

Nature is seen as another resource for, or factor of, production, similar to labor and capital. Subscribers to this discourse believe that it is possible to use natural resources perpetually and make business without putting in risk future regenerations. Supporters of corporate responsibility argue that nature is an input for production that has to be managed properly to make feasible business at the long-term. In this vein, the Competitiveness National Plan of the Peruvian State (PeruCompite website, 2011) refers to nature as a resource and its sustainable use will be a result of business efficiency and growth. The Plan argues that avoiding environmental degradation and protecting natural capital is in the benefit of the national economy. So protecting the environment will provide Peru good opportunities in the world market.

Agents and their motives

Business NGOs are considered key agents. Business NGOs provide expertise, promote social and environmental good practices, and facilitate the implementation, management and monitoring of social responsibility standards and guidelines for larger companies and their small providers. Business NGOs aim to increase competitiveness of companies in order to meet requirements of international buyers and influence the domestic business community to endorse the discourse. Business NGOs identify themselves as specialized key partners of companies interested in adopting social responsibility principles. Besides companies, business NGOs closely work with other specialized actors, such as consultancy firms and certification bodies.

Larger companies are also considered important agents. According to Rizo-Patron (2006, interview) motivations of larger companies to adopt business social responsibility can be pressure from social groups and from their transnational mother companies. Moreover, social responsibility has become the channel for companies to reflect about their roles in society and their position regarding social and political issues. Cuneo (2007) highlights the role of businesses as agents of social change in Peru by illustrating a more central participation of business in social and environmental problems. As Cuneo (2005) stated 'business social responsibility is not necessarily an issue of money, but of attitude and consciousness. This is the reason why companies that have low revenues adopt this concept'. Caravedo (2011) highlights the ethical motivation and emphasizes the adoption of social responsibility not only by companies, but also by NGOs and national governmental agencies. During the last two decades Latin American and Peruvian larger companies organized in platforms, such as Forum Empresa and the Peru 2021 Board of Trustees, have claimed to be socially responsible (Peru 2021 website, 2010).

Small providers have limited agency. As part of larger companies' value chains, small providers are persuaded to adopt business social responsibility principles. The small providers' motivation is economic growth, competitiveness and access to new markets. Only the small providers engaged with larger companies are considered to have enough capacity in terms of formality, competiveness, growth and standardization to adopt business social responsibility. Nevertheless, small providers depend on the pressure of the company dominating the value chain to adopt social responsibility. This means that the commitment of large buyers, exporters and retailers with social responsibility is key to pull small providers in that direction.

Most small scale enterprises do not have agency at all. Isolated small scale enterprises from larger companies' value chains are seen as incapable to implement business social responsibility. Most small scale enterprises are seen as inefficient enterprises, whose access is limited to low profitable markets. Proponents of the discourse claim that most small scale enterprises provide low quality employment, have bad working conditions, do not meet governmental regulation, do not produce according to standards of global markets, and producing products of low quality and not fashionable (Rizo-Patron, 2008, interview). Surviving informal enterprises are seen as something anomalous, which violates the definition of business since they are incapable to make profit and to grow. In the words of Rizo-Patron (2008, interview) 'if small scale enterprises are not competitive they should be out of the market and do something else, not business. If they are violating basic human rights they should not be at the market'. According to the discourse proponents, most small scale enterprises need to gain capacities in basic managerial skills and good production practices rather than in business social responsibility. The promotion of social responsibility in small scale enterprises is seen as a waste of investment.

National government is an ineffective agent. Subscribers to the discourse have several views regarding the role of the national government. While some assert that companies cannot replace government duties, others would claim that companies should take over. Regarding the former position, Forum Empresa (2009) states in a position paper: 'companies should take the lead in transforming the way of doing business to generate more profit for investors and more gains for the communities in which they operate'. Additionally, Peinado-Vara and De la Garza (2007) highlight the role of global value chains to increase the competitiveness of SMEs but also emphasize the role of government in leading reforms to formalize small scale enterprises. National government has the responsibility to support formalization and capacity building of small scale enterprises, especially ensuring the surviving of small scale enterprises (Peinado-Vara, 2007). Regarding the latter position, Cox (2006) asserts that 'the role of business seams gaining scope. The state is not capable to provide solutions for social problems. Then, this is an opportunity to stream private solutions to public problems'. By adopting social responsibility standards, companies aim to improve relations with their stakeholders, including the national government (Dias, Filomeno & Rizo-Patron, 2007). According to the discourse proponents, the fact that the national government focuses on supporting larger companies instead of SMEs is because it seeks short-term benefits. Large enterprises contribute more taxes and are less in number than small scale enterprises, so the first ones are easy to regulate. However, subscribers to this discourse point out that the cost of not regulating small scale enterprises is high in terms of, for instance, informality, accidents and the low quality of employment.

Key metaphors and their rhetorical devices

Key rhetorical devices in the discourse of corporate responsibility are the term 'competitiveness', 'stakeholders', 'small providers', 'codes of conduct', 'social balance', 'sustainability' and 'social and environmental standards'. Any report or speech of a CEO cannot miss the words competitiveness and stakeholders. The rest of the rhetorical devices are also common in the socially responsible business jargon.

6.3.4 Conclusions

a. Based on empirical research of business social responsibility in Peru, the discourse analysis identified two discourses: the business upgrading discourse and the corporate responsibility discourse. The worldwide spreading of social responsibility discourses is either forcing or encouraging larger companies, small scale enterprises, NGOs and the national government to gain new capacities to adapt to the new conditions. First of all, the study shows that business social responsibility is not a single body of thought but a space for diverse perspectives. Each discourse encompasses a particular storyline regarding business social responsibility and its application for SMEs. However, the two discourses share commonalities and differences in their views. The communalities are that both discourses basically involve the same entities and actors and the overall storylines rest on competitiveness and sustainability of business in the long-them. Both discourses consider the role of sustainable value chain to be central to improve the situation of small scale enterprises. The difference between both discourses is in the focus. While the business upgrading discourse focuses on social and environmental improvement of small scale enterprises *vis-à-vis* their economic growth, the corporate responsibility discourse focuses on building a good long-term relationship between larger companies and their stakeholders in a context of increasing market pressure.

b. The business upgrading discourse and the corporate responsibility discourse endorse the liberal market. The liberal market is understood as an entity connecting national and international economic actors in value chains at global level. In contrast, local markets are considered to have limitations in coping with the growth of domestic companies. Becoming socially and environmentally responsible is the only feasible way to legitimatize companies in society within the liberal market frame. Both discourses agree that the liberal market establishes favorable conditions to incorporate voluntary social and environmental standards, guidelines and ethical codes in SMEs. Proponents of the business upgrading discourse call for the need to connect SMEs as suppliers with socially responsible larger company value chains. Discourse proponents claim this is the only way that SMEs can survive global market pressure. Without upgrading, SMEs will not be able to cope with highly standardized production and sophisticated market and trading rules and requirements. Furthermore, without the connection to global supply chains it is impossible for small scale enterprises to grow. Proponents of the corporate responsibility discourse highlight the fact that social and environmental standards, guidelines and codes are not applicable for most small scale enterprises. They claim that small scale enterprises need to first meet basic business management capacities and to accomplish governmental regulation.

c. The discourses of business social responsibility consider nature as input of raw material for production. Environment is understood only in economic terms. Nature is another resource for production similar to labor and capital. It has to be protected and used in an efficient way since it warrants the endless cycle of production and consumption. Without raw material there is no business. The term 'environment' is most frequently used to refer to good conditions for business competitiveness such as safety, occupational health and risk prevention. While the business upgrading discourse emphasizes in social and environmental protection as factor to improve welfare of workers within small scale enterprises, the corporate responsibility discourse considers environment as a factor to keep their market share or expand it. The emphasis of the latter discourse on the economic perspective of the environment, narrows sustainability to only economic terms.

d. The business upgrading discourse and the corporate responsibility discourse identify common rhetorical devices such as 'social and environmental standards', 'profitable markets' and 'small providers'. The discourses also identify different metaphors. The business upgrading discourse has as central metaphors the words 'human rights' and 'vulnerable social groups'. 'Competitiveness' and 'stakeholders' are common in the corporate responsibility discourse jargon.

6.4 Discourses of business social responsibility in a two-dimensional policy realm

A two dimensional axis has been constructed to allocate the two discourses on business social responsibility (Figure 6.2). On the vertical axis (A) the focus is on the social foundation for structuring society, whereas in the horizontal axis (B) the subject is the economic foundation for organizing society.

The extreme poles of axis A represent 'individualistic' and 'communitarian' values. 'Individualistic' values highlight individual interests over collective interests and a faith in the capacity of individual action and ambition to create wealth and to bring about progress. The 'communitarian' values highlights the collective values and common identity of people within a shared geographical space from local to global community. In the first case, individual interests are the basis for relationships among actors. This pole refers to individual or group benefits, for instance of NGOs, producers, governmental agencies and consumers, rather than to collective welfare of the whole society. In the second case, the social strengthening of local communities is central, for instance of villages, small towns and cities. The collective values are, among others, solidarity towards people in need and a sense of community and cultural identity.

The extreme poles of axis B represent 'rights' and 'liberal market' values. 'Rights' values refer to the fairness of all human being based on social, economic and cultural justice. 'Liberal market' values refer to the mainstream the liberal market economy where business has a central role in production and trade of goods and services at the global level. In the first case, protection and defense of social and economic rights of 'disadvantaged' people against powerful actors are highlighted. Abuse might come from companies, transnational corporations and the national government. In the second case, giving the worldwide spreading of the liberal market and globalization, actors are interlinked globally. At the liberal market actors from local to global

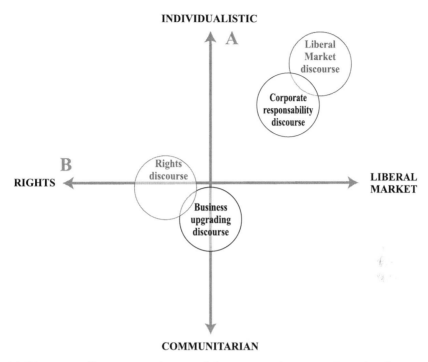

Figure 6.2. Discourses of business social responsibility in a two-dimensional social and economic axis.

scale are connected and global mobilization of goods and services are the basis for structuring economy in society.

The business social responsibility discourses are allocated in the two dimensional matrix (Figure 6.2). The business upgrading discourse is allocated at the crossing of 'communitarian' and 'liberal market' values. The corporate responsibility discourse is allocated at the crossing of 'individualistic' and 'liberal market' values. The rights discourse roots strongly on 'rights' values and links 'communitarian' and 'individualistic' values. While the upgrading discourse has its origin in the rights discourse, the corporate responsibility discourse has diverged from the liberal market discourse.

6.5 Major patterns and trends: networks and discourses

In this section networks and discourses are put together in a single framework (Figure 6.3) to understand the main patterns and trends. Five main patterns can be identified from the relationships between networks and discourses of business social responsibility: (1) two disconnected networks and two different discourses; (2) the rise of new actors; (3) business social responsibility discourses have evolved from adversarial discourses; (4) business social responsibility discourses establish good conditions for cooperation between companies and NGOs; and (5) the emergence of new roles of NGOs.

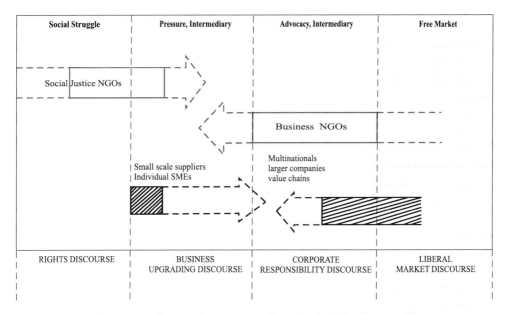

Figure 6.3. Main discourse and network patterns and trends of NGOs, SMEs and larger companies.

Regarding the first pattern, each network is organized around a particular discourse. The social justice network corresponds with the business upgrading discourse and the business network corresponds with the corporate responsibility discourse. Divergences in the discourses avoid the creation of connections between the social justice network and the business network. As a result, cooperation is limited to the actors within each network, especially at the local and regional level. This divergence in views between the business upgrading discourse and the corporate responsibility discourse is behind the establishment of Red Puentes Internacional as a counterweight to Forum Empresa and FUNDES. This means that social justice NGOs and business NGOs do not have intensive connections. The business network closely work with companies affiliated to the Peru 2021 Board of Trustees and the social justice network closely work with emerging small scale garment enterprises, agri-industries, shoemakers and handicraft makers. The social justice network is closer to the small scale enterprises embedded in a different social and market segment than the ones connected to the business network. Thus, the main difference between the two networks of business social responsibility is whether or not the beneficiaries are linked to global markets. At the global level, commonalities weight more than differences and bring common ground for dialogue among key global agents. However, at the regional and the local level, actors of both networks are in continuous clashes and mismatches despite the commonalities in their discourses.

The second pattern shows that a new type of NGOs – business NGOs – has emerged, to help companies to implement ethical codes, standards and monitoring the social performance of companies. The rise of business NGOs has been encouraged by larger companies to fulfill their needs and 'block' social justice NGOs. Social justice NGOs are conventional NGOs with a longer history that have 'harmonized' the right discourse with the business social responsibility core views. The rise of business NGOs is founded in the disagreement with the core viewpoints of the business upgrading discourse. This difference means that while social justice NGOs target small scale enterprises, business NGOs target larger companies.

Regarding the third pattern, NGOs that nowadays endorse business social responsibility used to endorse the right discourse or the liberal market discourse in the past. Social justice NGOs have adapted their 'rights' perspectives to the social responsibility perspective. Business NGOs have originated in the liberal market discourse and end up in the corporate responsibility discourse. However, social justice NGOs keep their core principles of rights and social justice, though abandon adversarial perspectives.

Fourth, NGOs and companies are starting to look at each other as potential partners rather than enemies. Before the 1990s it was common that NGOs identified companies as the enemy, not leaving any room for dialogue and cooperation. Currently, NGOs are adapting to the new global context characterized by the mainstreaming of liberal market ideas and the raise of small scale enterprises. This forces NGOs to catch up with standards, norms and other instruments of management. Particularly, conventional NGOs intend to change their 'adversarial' image towards companies by helping larger companies, small suppliers and potential suppliers to meet social and environmental standards and regulations. However, conventional NGOs are not abandoning pressure and checking strategies. Social justice NGOs are re-interpreting their traditional perspectives in favor of human rights and social justice under the current context of a globalizing liberal market. Companies are also changing towards cooperation, either forced by pressure from their stakeholders or motivated by the globalizing liberal market. However this process is in an early stage. What has been observed is that most of the time companies understand social responsibility in terms of social marketing rather than a change in business culture. Furthermore, companies prefer to work with business NGOs to implement social and environmental codes, guidelines and standards rather than with social justice NGOs. NGOs and SMEs increasingly become attached to global supply chains. Further pressure of NGOs and of other stakeholders to improve the social and environmental performance of SMEs is likely to happen.

Finally, NGOs are intermediaries between larger companies and their small suppliers, and between international cooperation agencies and SMEs. NGOs provide knowledge and advice for value chains dominated by larger companies on the implementation of standards, codes and guidelines of social responsibility. NGOs also implement projects about social responsibility in SMEs on behalf of international cooperation agencies. Those new roles of NGOs have emerged as an answer to the need of SMEs for accessing new markets and becoming suppliers of larger buyers, and the need of larger buyers to improve the social and environmental performance along the whole supply chain. So, NGOs and companies have adapted their traditional roles to the new context of a globalizing liberal market.

6.6 Conclusions

a. Although NGOs continue with pressure and advocacy as strategies to influence laggard companies towards change, they are also becoming intermediaries between larger companies and their small suppliers. This means that currently NGOs perform not only 'watchdog' roles but also 'helper' roles. In the traditional watchdog roles, NGOs have been involved in adversarial struggle with companies. In the new role of helper, NGOs are becoming new partners of companies. Business social responsibility networks involve not only conventional NGOs as key actor but also new type of NGOs. Conventional NGOs identified are social justice NGOs and the new type of NGOs identified are business NGOs. Social justice NGOs usually combine watchdog and helper roles in cooperating with larger companies, but business NGOs usually only perform helper roles. Performing intermediary roles is allowing NGOs to establish ties with economic and social actors and gaining more understanding of the market and social dynamics. This new scenario is fertile ground for cooperation between larger buyers and their small scale suppliers. However, being intermediaries represent not only opportunities but also risks for the legitimacy of NGOs. Social justice NGOs are facing a dilemma of how much to be engaged with and how much to keep distance from larger companies. Business NGOs have a clear position in their dog helper role.

b. The fact that companies are increasingly playing roles in the social sphere and NGOs in the economic sphere is related to the blurring of economic and social spheres in the networks and discourses of business social responsibility. Usually the economic sphere is seen as a reserved sphere for companies. Similarly, the social sphere is seen as being a monopoly for NGOs. In this process of changing roles, small scale enterprises perform in both spheres and are the mutual stakeholder for NGOs and larger companies. Small scale enterprises perform not only in the economic sphere but they are also part of local social networks. As small scale enterprises are getting connected with global markers, they connect their local social networks with global economic markets contributing to the further blurring of global social and economic spheres.

c. Conventional NGOs subscribing to business social responsibility discourses have abandoned adversarial perspectives, but their core claims of social justice have been transferred to the market arena. Abandon confrontation makes this type of NGOs not only less confrontational and less ideologically driven, but also more cooperative and technocratic. This fact particularly applies for social justice NGOs that are 'flirting' with larger companies to become a partner in improving the company's social and environmental downsides. In contrast, business NGOs were born in the economic sphere. Therefore, they are the preferred partner of larger companies. This fact makes business NGOs 'arrogant' with SMEs by considering them incompetent to adopt business social responsibility standards. It is worth mentioning that even though social justice NGOs have abandoned their adversarial perspectives, they usually are pointed at as confrontational actors. So, business social responsibility networks are not free of discrepancies, making cooperation between the social justice network and the business network hard. Social justice NGOs and business NGOs are strongly covered by mistrust shadowing the strong communalities between them. Adversarial views between companies and NGOs, and between social justice NGOs and business NGOs are 'phantoms' that contribute to create mistrust rather than to construct common ground for dialogue. Mistrust between the networks of business social responsibility shadows the strong commonalities between both discourses. In this context, SMEs constitute a valuable 'hub' between the two networks and discourses, and their further inclusion in the networks might encourage cooperation.

d. Competitive value chain suppliers are the target of business social responsibility agents. This means that most SMEs, especially small and micro scale enterprises, oriented to local markets are excluded from the networks of business social responsibility. There is no room for SMEs to gain power positions as they depend on their buyers. The business social responsibility discourses have prejudges towards small and micro scale enterprises as these are considered 'handicapped' enterprises. This underestimation of SMEs means that the agenda of the key network actors usually does not fit with the motivations of most SMEs. Networks and most SMEs have different interests. SMEs might potentially endorse social responsibility discourses but their viewpoints are not acknowledged by the key network actors. Without the participation of SMEs in the decision making of platforms, the business social responsibility discourses will not represent a feasible option to improve truly the social and environmental conditions of SMEs.

Chapter 7.
Sustainable production

7.1 Introduction

Most of the current efforts for making production sustainable in Latin America are concentrated in larger enterprises. A large number of small scale industries have been left relatively unnoticed, despite of their high resource intensity, inefficiency and high level of pollution load per unit of production (IDB website, 2011). This is so because of larger enterprises and SMEs in the region operate in a polarized business environment. First, productivity gaps between SMEs and larger companies in the region are more pronounced than in advanced countries, making it difficult for SMEs to establish commercial relations. Second, as a direct result of the first characteristic, Latin American SMEs are more isolated, less specialized and find it more difficult to join global value chains. Last but not least, companies in general, and SMEs in particular, have a higher degree of informality in Latin America (IDB website, 2011).

Sustainable production is currently applied in a wider range of SMEs, including industries, workshops and producer associations. The adoption of sustainable production by SMEs is not free of constrains. The lack of access to financial and technical resources and to capacity building has proven to be a critical bottleneck for the adoption of sustainable practices in SMEs (REDPYCS website, 2010).

The five more well-known approaches of sustainable production in Peru are 'cleaner technology', 'appropriate technology', 'eco-efficiency', 'technological innovation' and 'cleaner production'. Cleaner technology refers to economically competitive and productive technology that uses less material and energy, and generates less waste. Clean technology includes among others green building, green transportation, biofuels and natural gas. The current challenges of energy price spikes, resource shortages, greenhouse gas emissions and environmental security threats are contributing to make clean technologies appealing for industries (CIF website, 2012; IFC website, 2012). Appropriate technology evolves or is developed in response to a particular set of needs and in accordance with prevailing circumstances. Appropriate technology encompasses technological choice and was introduced in Peru by Practical Action International of UK to promote sustainable energy and resource use in rural small and micro scale enterprises and communities (Soluciones Practicas website, 2011). Eco-efficiency is a term coined by the World Business Council for Sustainable Development (WBCSD) in 1992. Eco-efficiency means the production of economically valuable goods and services while reducing the ecological impacts per unit of production. The reduction in ecological impacts translates into an increase in resource productivity, which in turn can create competitive advantage. Critical aspects of eco-efficiency include reduction in material and energy use, reduction of toxic materials, improving recyclability and use of renewable resources (IISD website, 2012). Technology innovation is the process through which new or improved technologies are developed and brought into widespread use. However, new technologies might create or facilitate increased pollution, or may mitigate or replace existing pollution activities. Therefore, policy makers have to consider the environmental and social consequences of technological innovation (Jaffe, Newell & Stavins, 2003; 2020 Science website, 2012). Finally, cleaner production constitutes one of the relevant preventive strategies to tackle environmental problems in the processing of products and services (UNIDO, 2010). Cleaner production was introduced in Peru by the United States Agency for International Development (USAID) in the early 1990s for the fish industry (Cleaner Production website, 2010).

Sustainable production involves in Peru a variety of NGOs. NGOs handling a specialized 'know-how' on technology, which – I called 'technocratic' NGOs for purposes of this research, are part of the networks and discourses promoting sustainable production. Although technocratic NGOs appeared in the 1970s and 1980s, most of them matured around the 1990s onwards. The mainstreaming of the liberal market in the national economy and the need of specialized knowledge and information by SMEs are factors contributing to the rise of technocratic NGOs. Most technocratic NGOs in sustainable production networks have expertise on eco-efficiency and appropriate technology, and are oriented to urban and rural SMEs and micro-enterprises. Technocratic NGOs work as consultancy organizations and provide services to SMEs on sustainable production. Technocratic NGOs are different from conventional NGOs in the sense that conventional NGOs do not have specialized expertise on technology. Governmental and inter governmental agencies operating projects on sustainable production, usually with funding from international cooperation agencies are considered government organized non-governmental organizations (GONGOs) for the purpose of this research. Technocratic NGOs and (inter)governmental agencies (GONGOs) are usually led by committed professionals and experts formally educated at universities. Examples of technocratic NGOs in Peru are among others CER, Practical Action Peru and IPES. Examples of (inter)governmental agencies in Peru are among others CONCYTEC and ITACAB.

The chapter is organized in six sections. Section 7.2 presents the network analysis of sustainable production and describes in detail the main networks identified in Peru: the eco-efficiency network, the appropriate technology network, the cleaner technology network, the technological innovation network and the urban cleaner production network. Afterwards, the main patterns, challenges and trends of the networks, and the main conclusions of the network analysis are presented. Section 7.3 presents the discourse analysis of sustainable production and describes in detail the main discourses identified in Peru: the discourse of cleaner production and the discourse of appropriate technology. Subsequently, the main conclusions of the discourse analysis are presented. In Section 7.4 the two discourses on sustainable production are allocated in a two dimensional policy realm. Section 7.5 presents the major patterns and trends of the networks and the discourses of sustainable production. Finally, the main conclusions of the chapter are presented in Section 7.6.

7.2 Network analysis of sustainable production

The analysis of networks focuses on identifying the following issues: the key NGOs involved in the network of sustainable production fostering sustainable SMEs, the network relationships in terms of cooperation, the resource exchange and power relationships among network members, and the main network changes and challenges. The analysis concentrates on the sustainable production supporting SMEs in Peru and their connections with other branches at regional and global level.

Before analyzing them in depth, the five main networks involved in the promotion of sustainable production for SMEs in Peru are briefly introduced. Each network includes a particular set of national, regional and global actors.

The first network to be analyzed is the eco-efficiency network. The key NGO in the eco-efficiency network in Peru is the Eco-efficiency and Social Responsibility Center (Centro de Ecoeficiencia y Responsabilidad Social, CER in Spanish acronym). At the regional level, CER is affiliated to the Latin American Platform of the National Cleaner Production Centers. The second network to be analyzed is the appropriate technology network. The key NGO identified in this network is the Institute for the Transfer of Technology for Marginal Sectors (Instituto de Transferencia de Tecnologías Apropiadas para Sectores Marginales, ITACAB in Spanish acronym). At the regional level, ITACAB is affiliated to the cleaner production regional platform of the Andres Bello Intergovernmental Treaty (CAB). ITACAB is also the coordinator of this regional platform. The third network is the cleaner technology network, particularly the Peruvian branch. The key actor in this network is the National Council of Science and Technology (Consejo Nacional de Ciencia y Tecnología, CONCYTEC), a governmental agency. At international level, CONCYTEC is affiliated to the technological transfer platform (TTN). The fourth network is the technological innovation network. The key actors in this network are a cluster of private and public organizations named Centers of Technological Innovation (Centros de Innovación Tecnológica, CITEs in Spanish acronym) and their central coordination office. Finally, the urban cleaner production network is analyzed. This network includes a single NGO named the Peruvian Institute of Social Economy (IPES in Spanish acronym) connected to several international cooperation agencies promoting cleaner production in urban small scale enterprises. In delivering support to SMEs within their respective networks, CER, ITACAB, CONCYTEC, CITEs and IPES collaborate with other NGOs, SMEs associations, municipalities, governmental agencies and international cooperation agencies.

7.2.1 The eco-efficiency network

Key actors: Eco-efficiency and Social Responsibility Center

The Eco-efficiency and Social Responsibility Center (Centro de Ecoeficiencia y Responsabilidad Social, CER in Spanish acronym) was launched in 2002 with the initial name of Center of technological efficiency (Centro de Eficiencia Tecnológica, CET Peru in Spanish acronym). In 2007 the center changed its name to CER. In this new phase, CER is not an independent organization anymore but is part of the Group of Environmental Entrepreneurs (Grupo GEA in Spanish acronym) (Table 7.1). The Grupo GEA is a Peruvian Lima-based NGO (Espinosa, 2008, interview).

CER aims to promote competitiveness of national companies by implementing eco-efficiency measures and social responsibility standards, training consultants and providing advice to policy makers. CER focuses on diagnosis, elaboration of eco-efficiency plans, development of indicators, implementation and monitoring of the efficient use of natural resources and raw materials of production, and the reduction of waste (CER website, 2011; Espinosa, 2008, interview). CER is also the technical guarantor of the Peru Credito bank for the upgrading of cleaner technology in SMEs and evaluates the Clean Development Mechanism applied to eco-efficiency projects of SMEs in Peru. The Clean Development Mechanism is one of the flexible financial mechanisms

Table 7.1. Key national and international partners, and allied organizations of CER.

	Organizations
Peru	Grupo GEA
	Regional government of Callao
	Ministry of Production
	Ministry of Environment
	Peru 2021
International	Swiss Agency for Development and Cooperation (SDC)
	State Secretariat for Economic Affairs of the Swiss – Federation (SECO)
	United Nations Industrial Development Organization (UNIDO)
	United States Agency for International Development (USAID)

Source: Espinosa, 2008, interview.

of the Kyoto Protocol that aims to compensate CO_2 emissions of an industrialized country by financing projects of CO_2 reduction in developing countries. The work of CER is mainly oriented to larger and medium-sized companies, most of them located in Lima (CER website, 2011; Espinosa, 2008, interview).

CER currently (2012) implements three projects: EcoADEX, EcoHotels, and EcoParks. The EcoADEX project focuses on promoting cleaner production in exporting SMEs. The EcoHoteles project focuses on improving eco-efficiency and reducing carbon food-print in their affiliated hotels. The EcoPark project aims to reduce operational costs and improve environmental performance in a set of industries located in the province of Callao-Lima, as 25% of the country's production is located in this province. Callao concentrates 40% of the national industry and 66% of the manufacturing industry with high levels of industrial pollution (Cleaner Production website, 2010). The Ecopark project is a pioneer experience of cleaner production in Peru that consider geographical considerations in terms of applying eco-efficiency measures to companies that are located in a same geographical area (Alegre and Marthaler, 2010).

Key actors: the Latin American Cleaner Production Platform

The Latin American Cleaner Production Platform was launched in 2005 with participation of 12 countries of the region: Bolivia, Brazil, Colombia, Costa Rica, Cuba, Ecuador, El Salvador, Guatemala, Honduras, Mexico, Nicaragua and Peru (Table 7.2). The Mexican Center of Cleaner Production (Centro Mexicano para la Producción más Limpia, in Spanish) is the current coordinator of the platform. The platform aims to enhance cooperation and exchange of good practices on environmental sound technologies among the 12 National Cleaner Production Centers (NCPCs). Through a site on internet (www.produccionmaslimpia-la.net), the platform disseminates information about national and international events, guidebooks and technical reference manuals. Moreover, the platform facilitates the sharing of cases of application of cleaner

Table 7.2. Affiliates and partners of the Latin American Cleaner Production Platform.

	Latin American Cleaner Production Platform
Affiliated organizations	Centro Ecuatoriano de Producción Más Limpia
	Centro Mexicano Para la Producción Más Limpia
	Centro Nacional de Producción Más Limpia Colombia
	Centro Nacional de Producción Más Limpia Costa Rica
	Centro Nacional de Producción Más Limpia de Honduras
	Centro Nacional de Tecnologias Limpas SENAI Brasil
	Centro de Ecoeficiencia y Responsabilidad Social – CER
	Centro de Ingeniería Genética y Biotecnología de Cuba (CIGB)
	Centro de Producción más Limpia de Nicaragua
	Centro de Promoción de Tecnologías Sostenibles CPTS Bolivia
	Cuba – Agencia de Medio Ambiente (AMA)
	Fundación Centro Guatemalteco de Producción más Limpia
	Fundación Centro Nacional de Producción Más Limpia de El Salvador
	Instituto Cubano de Investigaciones de los Derivados de la Caña de Azúcar (ICIDCA)
	Instituto de Investigaciones en Fruticultura Tropical IIFT
	Instituto de Investigaciones para la Industria Alimenticia (IIIA)
International cooperation agency	State Secretariat for Economic Affairs of the Swiss Federation (SECO)
Inter-governmental agency	United Nations Industrial Development Organization (UNIDO)

production, of documents about environmental legislation of each country member of the platform, of monthly reports and of links on cleaner production (Cortijo, 2006; CPLatinNet website, 2011; Espinosa, 2008, interview). The management of the platform is structured in five hierarchical levels: executive board, advisory committee, Programme coordinator, national focal points and working groups on specific topics (CPLatinNet website, 2011). This structure is for purposes of coordination with the United Nations Industrial Development Organization (UNIDO) and facilitates learning and exchange among NCPCs.

NCPCs have the aim to build capacities and transfer knowledge on environmental sound technologies, provide consultancy advises to companies, raise awareness, and support the development of governmental policies on sustainable production. UNIDO and the United Nations Environmental Programme (UNEP) launched in 1994 a joint programme to establish NCPCs all over the world with the financial support of the governments of Switzerland and Austria. In supporting NCPCs, UNIDO aims to embed international environmental policies, especially regarding the Millennium Development Goals and the Clean Development Mechanism into local SMEs (UNIDO website, 2011).

Based on the lessons and experiences of NCPCs, UNIDO and UNEP have joined efforts to establish the Joint UNIDO-UNEP Programme on Resource Efficiency and Cleaner Production

(RECP). RECP is the application of an integrated preventive environmental strategy to processes, products and services to increase efficiency and reduce risks to humans and environment. RECP is the new global platform launched in 2010 to be the global 'umbrella' for NCPCs worldwide (UNIDO website, 2011; UNEP website, 2011).

Network relationships: cooperation

From 2002 to 2007, CER cooperated closely with several national organizations working on cleaner production, including the Lima University, the Agrarian University, the National Society of Industry and the National School for Industrial Training. At this time, cooperation aimed to promote eco-efficiency and the improvement of working conditions as strategies to promote competitiveness of SMEs in Peru (GEA website, 2011). During that period, CET was affiliated to the CAB Regional Platform of cleaner and organic production (Cortijo, 2006, interview). From 2007 onwards, CET starts a new phase with the name of CER; expanding its cooperation ties to new governmental and business actors to reach more SMEs (GEA website, 2011).

Regarding cooperation with NGOs, CER works closely with the Grupo GEA. In 2007, CER became the branch of Grupo GEA dealing with sustainable development of SMEs. The Grupo GEA establishes the overall institutional policy of CER for their development activities. CER also works closely with Peru 2021. While CER emphasizes technology and production processes, Peru 2021 emphasizes management standards. Although most SME clients of CER are located in Lima, CER is also reaching SMEs located in other parts of Peru by establishing partnerships. CER has ties with the Service Center for Businesses of Arequipa (CECEM). CECEM is an organization that provides services on cleaner production and business networking in the southern part of the country (Alegre, 2011, interview). Additionally, CER cooperates with the Programme of Competitiveness, Innovation and Development of the Arequipa region (CID). CID is a NGO also based on Arequipa with its focus on the south of Peru, which aims to improve business competitiveness and social responsibility by engaging small scale enterprises in export chains (Espinosa, 2008, interview). At regional level, CER is connected to the 12 centers of cleaner production of Latin America (CER website, 2011; Espinosa, 2008, interview).

Regarding the cooperation with business associations, CER works closely with Peru's exporters association (Asociación de Exportadores, ADEX). The relationship with ADEX started in 2005 and the cooperation has been central to promote eco-efficiency in exporting SMEs to reduce operational costs and making production carbon neutral. ADEX and CER launched EcoADEX in 2009 (ADEX website, 2011; Burga, 2009). CER has also partnered with the Hotels Society of Peru (Sociedad Hoteles del Perú, SHP) and ADEX to implement the project EcoHoteles. The EcoHoteles project allowed the reduction of more than 10% of GHG emissions, energy consumption reduction of 15%, fuel reduction of 10% and water consumption reduction of 35%, with recovering investments in less than a year (Burga, 2009). Cleaner production measures have been implemented in the following five star hotels: Los Delfines, Miraflores Park Hotel, Casa Andina and Inka Terra lodge. Also, CER has measured the CO_2 emissions of the Inter-American Development Bank (IDB) office in Peru and developed a plan for IDB office Peru to compensate its carbon foot-print by financing projects on eco-efficiency in Peru. The aim of CER in 2010 was to implement eco-efficiency and social responsibility measures in 3 services companies, 4 fishmeal industries, 10 hotels and 15 manufacturing industries (Alegre, 2009; CER website, 2011). During 2008-2009 under the Ecopark project, 7 well-known companies (Alicorp, Calsa, Zinsa, Silgelsa, Texfina, Cormin Callao and Corporación Rey) located in the province of Callao implemented eco-efficiency measures in the production processes (GEA website, 2011). The cooperation of the Industrial Park of Arequipa Associations (ADEPIA), next to the aforementioned Service Center for Businesses of Arequipa, has contributed to establish an Ecopark in Arequipa (Alegre, 2011, interview).

Regarding the cooperation with SMEs, CER provides consultancies for SMEs. Most of their clients are SME providers to larger domestic companies and exporters. CER has implemented eco-efficiency measures in several manufacturing companies, particularly textile and furniture and agri-industries, including asparagus producers and slaughter houses. Over the period 2002 to 2012 CER has implemented cleaner production measures in several SMEs, including Arquitecma Peru SAC, Nena's Collection, Metalexacto, Mercurio Industria y Comercio SAC, Fabrica Textil La Bellota SA, Textiles San Sebastián SAC, tannery La Pisqueña S.A, tannery Avidas SRL, Maderas Peruanas SA and Papelera Panamericana (CER, 2007; Febres, 2004; Hechaverria, 2004; IPES website, 2012; Sibille, 2004). CER has also measured carbon foot-prints in several SMEs to improve production performance and reduce GHG emissions, for example in Topy Top, Corporación Rey, Texfina (textile), Agrícola ATHOS (agri-industry), Calsa – Fleshman (bakery products industry), Cormín Callao (mineral storages), Silgelsa (chemicals), Zinsa (fundring) and COPEINCA (fishing).

Regarding cooperation with national banks, CER has provided advice to introduce environmental friendly production technology in local SMEs. CER has ties with the following banks: Peru Credito Bank, Scotiabank and Interbank. Several SMEs have also accessed environmental credits, getting reimbursement up to 40% of their total investment (for instance, Agropecuaria Esmeralda, Alianza Metalúrgica, CURPISCO, Helbert Samalvides Dongo, Molino Arrocero El Corso, MAPESA and Papelera Panamericana) (CER website, 2011; GEA website, 2011; Espinosa, 2008, interview).

Regarding cooperation with national governmental agencies, CER provides expertise to the Ministry of Environment of Peru's eco-efficiency Programme. This programme aims to promote eco-efficiency in public administration, municipalities, schools and companies. The Ministry of Environment of Peru is the national government representative in the EcoADEX project. The Ministry of Environment of Peru, the Ministry of Production (PRODUCE, in Spanish acronym) and the Regional Government of Callao (GORE Callao) provide support to the EcoPark project in the province of Callao. In 2010 with the advice of CER and the support the regional government of Pucallpa, a new eco-industrial park has been implemented in Pucallpa (Alegre, 2009; CER website, 2011; GEA website, 2011; Espinosa, 2008, interview). This new project is being implemented in the region of Pucallpa in cooperation with the small forest wood business association (APEMEPD) (CPLatinNet website, 2011). Finally, in cooperation with the National Direction of Environmental Health (Dirección General de Salud Ambiental of Peru, DIGESA), CER operates the 'Quimicos Seguros' project, aimed to develop a regulative framework for the management of chemicals in Peru.

Regarding the collaboration with international cooperation agencies, CER receives support from UNIDO. UNIDO provides policy guidance and expertise to CER's overall strategic planning and to the EcoADEX, EcoHotels, EcoParks and 'Quimicos Seguros' projects (CER website, 2011; Espinosa, 2008, interview). UNIDO established the national cleaner production center of Peru under the name of Eco-efficiency and Social Responsibility Center (CET) in 2002 with the financial support of the United States Agency for International Development and the State Secretariat for Economic Affairs of the Swiss Federation (SECO). In 2010 the United States Agency for International Development also provided financial support to the eco-industrial park implemented in Pucallpa (CPLatinNet website, 2011). CER gets financial support from SECO under a bilateral agreement between the governments of Switzerland and Peru (Espinosa, 2008, interview). SECO also provides funding for the EcoPark project. The financial support of SECO is channeled to CER by the Swiss Agency for Development and Cooperation (SDC). SDC also acts as guarantor for environmental credits to upgrade old environmental unfriendly technologies within SMEs through national banks in Peru. UNIDO and SECO provide also support to the Latin American Cleaner Production Platform (CPLatinNet website, 2011; Cortijo, 2006, interview; Espinosa, 2008, interview). Additionally, two Swiss research institutions, the Swiss Federal Laboratories for Materials Testing and Research (EMPA) and Neosys, provide specialized technical advice to CER on cleaner production (CER website, 2011; Espinosa, 2008, interview). Moreover, CER has been part of the 'Quimicos Seguros' project of Peru. This project is promoted by UNIDO and UNEP and funded by the Strategic Approach to International Chemicals Management (SAICM) for two years (2009-2010) (CER website, 2011; SAICM Peru website, 2012). Finally, through close cooperation with UNIDO, CER is tied to international platforms on cleaner production. CER is part of the Latin American platform of National Cleaner Production Centers (Red Latinoamericana de Producción Más Limpia, in Spanish) and the worldwide NCPCs. It is important to point out that UNIDO and UNEP, as was mentioned above, have recently overcome duplication of efforts and strengthened cooperation by launching the RECP platform. Recently CER has joined this new platform of global reach (UNIDO website, 2011; UNEP website, 2011; Alegre, 2011, interview).

Network relationships: power

This section analyzes the power relationships in the eco-efficiency network, focusing on key organizations identified in Peru and their international connections. The power distribution along the network and their nodes at national, regional and global level are analyzed in depth.

CER is the national node, the Latin American Cleaner Production Platform is the regional node and UNIDO is the global node of the eco-efficiency network. While SECO provides the financial support and UNIDO designs the master plan for intervention in NCPCs, CER is the project operator at national level in Peru.

The relationship among the 12 Latin American NCPCs is horizontal in the platform. The platform encourages Latin American NCPCs to apply together for funding and provides them a better position to negotiate with funding agencies. However, the capability of NCPCs to come up with initiatives to the platform and work in cooperation has been hard to establish.

UNIDO has a long history promoting cleaner production in industries worldwide and influences strongly the eco-efficiency network. One of the outcomes of the World Summit for Sustainable Development in 2002 was to encourage UNEP to initiate and promote a ten-year programme framework on sustainable consumption and production, known as the Marrakech Process (UNIDO website, 2011; REDPYCS website, 2010). The formation and operationalization of the Latin American Cleaner Production Platform and NCPCs are related to this mandate. UNIDO provides the agenda to the network, designs projects and manages the operation worldwide of

NCPCs in 42 countries, including the Latin American NCPCs (Espinosa, 2008, interview). UNIDO has established the institutional infrastructure of NCPCs, mediating the funding allocation from donors to the Latin American platform. It also provides annual incentives to encourage cooperation of the platform's affiliated NGOs (UNIDO website, 2011)

UNEP is another global champion in the network, but is not as central as UNIDO. SECO and the United States Agency for International Development are the main agencies that provide the financial support to the Latin American platform and NCPCs in the region. The Latin American NCPCs work according to the agenda established by UNIDO and operate projects at national level. After the changes in administration in 2007, CER has also consolidated its position in the network as national node. The consolidation is related to the formation of new ties with powerful economic actors, such as the Peru's exporters association and the Hotels Society of Peru. These business associations are the gatekeepers to reach domestic SMEs engaged in global markets. CER has also strengthened ties with the Ministry of Environment and the Ministry of Production. In the last years CER is also building up ties with regional governments of Callao and Pucallpa.

The power position of UNIDO and UNEP is based on their status as UN agency and their capacity to influence donors. However, this power position does not match with the limited impact of cleaner production at national level. At national level, there is a gap between the aims of UNIDO/UNEP and of the NCPCs to expand cleaner production on the one hand, and the motivation of SMEs to adopt it on the other hand. This is shown by the strong financial dependence of the NCPCs and the scarce number of SMEs adopting cleaner production. This gap has been realized by UNIDO in an evaluation of the programme done in 2007, concluding that although NCPCs are succeeding in putting cleaner production on the agenda of businesses and government, the improvement of effectiveness and efficiency in the implementation of the programme and the financial independence of NCPCs are the main challenges (UNIDO website, 2011).

No relevant conflicts have been identified in the eco-efficiency network. This might be so for several reasons. First, CER has been appointed as a NCPC by a globally well reputed agency such as UNIDO. So, hardly any other organization can dispute this position of CER in the network. The support of UNIDO to CER is of tremendous help by channeling funding from SECO. Second, there are few NGOs with expertise in cleaner production in Peru to dispute the position of CER. Third, the market for services on cleaner production in Peru is small. CER is pioneering in creating the need of making production cleaner. There is no room for disputes and conflicts on power positions with other organizations since ties between similar NGOs are scarce. As the network evolves, potential disputes might come from the nodes of other networks dealing also with cleaner production and technology in SMEs.

7.2.2 The appropriate technology network

Key actors: the Institute for the Transfer of Technology for Marginal Sectors

The Institute for the Transfer of Technology for Marginal Sectors (Instituto de Transferencia de Tecnologías Apropiadas para Sectores Marginales, ITACAB in Spanish acronym) is an international organization established in 1975 to implement the Andres Bello Intergovernmental Treaty (CAB in Spanish acronym). CAB aims to promote science and technology, and cultural exchange among their members. CAB was signed in 1970 by 6 Latin American countries: Bolivia, Chile, Colombia, Ecuador, Peru and Venezuela. Later, the countries of Panama, Spain, Cuba, Paraguay, Mexico and Dominican Republic and Argentina joined the CAB (Convenioandresbello website, 2012; ITACAB website, 2011).

ITACAB aims to facilitate the knowledge exchange among the CAB country members, and to develop and transfer sound technologies for SMEs development, in order to improve the condition of marginal sectorsin society. ITACAB became operational in 1986 and works on themes such as agri-industry, agriculture, energy and small scale industries (ITACAB website, 2011; Sanchez, 2009, interview). The organization's headquarter is located in Lima, Peru (ITACAB website, 2011).

ITACAB leads the Center of Resources for the Transferring of Technologies (CRTT, in Spanish acronym). CRTT is an online database of appropriate technologies implemented in 2004. CRTT includes 312 case studies on appropriate technologies from SMEs, NGOs and governmental agencies of CAB countries. (Sanchez, 2010, interview).

Key actors: Practical Action

Practical Action is an international NGO founded in 1966 which works with low income communities to develop appropriate technologies in food production, agro-processing, energy, transport, water and sanitation, shelter, climate change adaptation and disaster risk reduction (Practical Action website, 2010). The target of Practical Action is farmers in rural areas, producer associations and small scale agri-industries. Practical Action was formerly known as Intermediate Technology Development Group Ltd (ITDG). Since September 2009 the new name of the organization is Practical Action (Practical Action website, 2010). The organization is under the direction of a Board of Trustees. The senior management team comprises the Chief Executive and twelve Directors, including seven Country and Regional Directors (Practical Action website, 2010).

Practical Action has 7 offices worldwide (Bangladesh, East Africa, Latin America, Nepal, South Asia, Southern Africa and Sudan) and a Head Office in the UK. The Latin American offices are located in Peru and in Bolivia. Practical Action works in Peru since 1975 in 9 regions (Ancash, Ica, Cajamarca, Junín, Apurímac, Huancavelica, Lambayeque, Cusco and San Martín). Practical Action Peru (Soluciones Prácticas Peru, in Spanish) promotes technologies such as small hydropower plants, solar panels, irrigation systems, farming techniques, sanitation, agro processing and drinkable water production in rural communities in Peru. Practical Action Peru mainly provides expertise in technology for farmers in rural areas and for small scale agri-industries aimed to widen their opportunities at markets (Practical Action website, 2010).

In its institutional policy for the period 2007-2015, Practical Action puts at the top of its agenda the development and allocation of technologies for climate change adaptation in rural areas, as well as influencing national public policy in those issues (Practical Action website, 2010).

Key actors: CAB Regional Platform of Cleaner Production

The CAB Platform of Cleaner Production (Red CAB de Producion Mas Limpia, in Spanish) aims to improve the competitiveness of small scale enterprises in order to fight poverty in the countries of the Andres Bello Treaty (CAB).

The CAB platform of cleaner production was established in 2002 with the support of the German Technical Cooperation (Deutsche Gesellschaft für technische Zusammenarbeit, GTZ). The CAB platform started to operate in 2003 under the coordination of ITACAB. Since 2006 the CAB platform of cleaner production does not implement projects anymore (Sanchez, 2009, interview) and it is just kept as an e-platform to exchange information (Vidal, 2006, interview). According to the information available till 2009 (CAB platform website, 2009) the CAB platform had a coordinator and a management committee integrated by each country member and the 37 organizations of 7 countries members of CAB (Table 7.3). After the financial support of the German Technical Cooperation ended, ITACAB stopped coordinating the CAB platform (Sanchez, 2010, interview) and the affiliated organizations started abandoning the platform. In 2010 the e-platform was canceled and currently (2012) it is not available at internet anymore.

Table 7.3. Key national and international partners of CAB platform in 2003.

	Organizations
Peru	Colegio de Ingenieros del Perú – Departamental Lima: Capítulo Ingeniería Pesquera
	Universidad Nacional de Ingeniería (UNI)
	Mesa de Salud y Seguridad en el Trabajo (OIT/ OPS)
	Pontificia Universidad Católica del Perú
	Pontificia Universidad Católica del Perú. Instituto de Estudios Ambientales (PUCP-IDEA)
	Centro de Eficiencia Tecnológica (CET Peru)
	Universidad Nacional de ingenieria – Centro Peruano Japonés de Inves. Sísmicas y Mitigación de Desastres
	Asociación de Exportadores (ADEX)
	Servicio Nacional de Adiestramiento en Trabajo Industrial (SENATI)
	Agencia Suiza para el Desarrollo y la Cooperación (COSUDE/ Programa de Capacitación Laboral (CAPLAB)
	Sociedad Nacional de Industrias (SNI)
	Instituto de Imagen de la Mujer, Joven, Empresario y Niño (IMAJEN)
	Universidad de Lima – Centro de Estudios Ambientales (CEA)
	Consejo Nacional de Ciencia y Tecnología (CONCYTEC)
	Instituto de Investigación y Capacitación Municipal (INICAM)
Colombia	Asociación Colombina de Pequeñas y Medianas Industrias (ACOPI)
	Departamento Técnico Administrativo de Medio Ambiente (DAMA)
	Ministerio de Ambiente, Vivienda y Desarrollo Territorial de Colombia
	Corporación para la Investigación Socioeconómica y Tecnológica de Colombia (CINSET)
Bolivia	Federación Boliviana de la Pequeña Industria (FEBOPI)
	Instituto de Asistencia Social,Economica y Tecnologica (INASET)

>>

Table 7.3. Continued.

	Organizations
Venezuela	Universidad Central de Venezuela – Centro de Estudios del Desarrollo
	Federación Venezolana de Cámaras de Comercio y Producción de Venezuela
	(FEDECAMARAS)
	Fondo Venezolano de Recoversión Industrial y Tecnológica (FONDOIN)
	VITALIS Gente, comunicación y ambiente
Chile	Conf. Gremial Nac. Unida de la Mediana, Pequeña, Microindustria, Servicios y
	Artesanado de Chile (CONUPIA)
	Fundación Chile
	Centro de Transferencia de Tecnologías Limpias (CTTL S.A.)
Ecuador	Escuela Politécnica Nacional-Centro de Estudios para la Comunidad
	Inst. de Invest. Ecuador
	Ministerio Del Ambiente
	Centro Ecuatoriano de Producción Más Limpia (CEPL)
	Cámara de la Pequeña Industria del Azuay Gremio Empresarial
Panama	CONEP – UNPYME
	Univ. de Panamá – Facultad de Ciencias Naturales Exactas y Tecnología (CEREB)
	Autoridad Nac. del Ambiente. Prog. de Instrumentos de Gestión Ambiental y
	Participación Empresarial en la Producción mas Limpia
	Red de Empresas en Producción Más Limpia

Network relationships: cooperation

Since its foundation ITACAB, Practical Action Peru and the CAB platform have worked in cooperation with NGOs, universities, national governmental agencies and international cooperation agencies to promote cleaner and appropriate technologies in SMEs.

During 2002-2006, ITACAB was very active linking national and regional actors involved in cleaner production. ITACAB has worked together with COPEME, Practical Action and other NGOs to improve production processes of small scale enterprises and small scale producers (Sanchez, 2009, interview). Since 2004, ITACAB has established ties with producers, small scale enterprises, rural development promoters, researchers, and students, through the Center for Technological Transferring of Resources (Sanchez, 2010, interview). During 2003-2006, ITACAB worked closely with the National Council of Science and Technology (CONCYTEC, see below) (Sanchez, 2009, interview). ITACAB also contributed with case studies about low cost technologies to the 'Sustainable Alternatives Network (SANET)'. SANET is an online database (http://www.sustainablealternatives.net) of cleaner and ecological technologies from several countries, including Peru. In Peru, CONCYTEC coordinates the SANET platform. ITACAB also has ties with Practical Action Peru and the Centers of Technological Innovation but they do not work together in transferring technology for SMEs. ITACAB has also strong ties with production-oriented, environmental protection-oriented and education-oriented governmental agencies of country members of the CAB. ITACAB has ties with the Andean Nations Community (CAN) to develop Andean regional policies on cleaner production. ITACAB has also ties with the Food and Agriculture Organization of the United Nations (FAO), the United Nations Development Programme (UNDP), the Commission for Latin America and the Caribbean (CEPAL), the German Technical Cooperation and the Spanish Agency for International Cooperation (AECI). These agencies provide ITACAB with financial support and policy guidance to their projects. The German Technical Cooperation has been the main funder of ITACAB and the CAB Regional Platform of Cleaner Production till 2006. Actually the only perational project of ITACAB in appropriate technology for SMEs is the online database Center of Resources for the Transferring of Technologies. This project is funded fully by CAB countries, not by donor agencies anymore. ITACAB is exploring funding from donor agencies for new projects in the field of appropriate technologies (Sanchez, 2010, interview).

Practical Action Peru works closely with their headquarter office in UK and their branches in several countries worldwide. At the national level, Practical Action Peru works in close cooperation with municipalities, small scale producers, small scale enterprises and other NGOs. Practical Action Peru had a very active role in the CAB Regional Platform of Cleaner Production till 2006. Practical Action and ITACAB have ties but currently they do not have collaborative work related to technology for small scale enterprises. Practical Action Peru has set up the Centre of Demonstration and Qualification in Appropriate Technologies (Centro de demostración y capacitación en tecnologías apropiadas, CEDECAP) in Cajamarca and the 'Kamayuq' Technical School in Puno. The schools provide training in renewable technologies and information and communication technologies (ICT) to set up small scale enterprises in rural areas. Moreover, Practical Action Peru has close ties with the Network of Rural Agri-industries of Peru (REDAR), a platform that groups 51 organizations nation wide. At international level, Practical Action Peru has cooperation among others with the European Commission, OXFAM America, Inter-American Development Bank, the Food and Agriculture Organization of the United Nations (FAO) and several international foundations (Soluciones Practicas Memoria Institucional 2010-2011, 2011). Taking the need of technology for climate change adaptation as umbrella concern, Practical Action International is building ties with a wider set of actors over the last years. During the last years Practical Action International is also building ties with business corporations, including Toyota Foundation and Northern Foods, international foundations, including WW Kellogg foundation and Zurich Foundation, and national companies, including EDEGEL, by collaborating in their business social responsibility programmes (Soluciones Practicas Memoria Institucional 2011-2012, 2012).

Network relationships: power

The appropriate technology network has currently nodes at national and global level. At the national level, the main node of the network is Practical Action Peru. At global level, the node is Practical Action International. The regional node is isolated from the national and global nodes. During the years 2002-2006, the CAB Regional Platform of Cleaner Production and ITACAB were regional nodes of the appropriate network (Sanchez, 2010, interview). Currently ITACAB does not perform as a node anymore. However, cooperation was not found between Practical Action Peru and ITACAB, nor between the nodes of the appropriate technology network and the nodes of the organic production networks. They do not work in coordination.

The nodes of the appropriate network are very 'flexible'. While ITACAB has moved backwards and the CAB Regional Platform of Cleaner Production was dismantled, Practical Action Peru has kept its position. ITACAB and Practical Action Peru have their own funders, partners and SME beneficiaries, especially after the failure of the CAB Regional Platform of Cleaner Production. ITACAB and Practical Action are parallel nodes in the network. However, while ITACAB has faced shortages in their operation budgets during the last years, Practical Action Peru is expanding their ties. As Practical Action Peru is part of Practical Action International, Practical Action Peru has more financial capacity, expertise and ties than ITACAB. For instance, Practical Action Peru has expanded their developmental activities to two more regions (Huancavelica and Lambayeque). Similarly, Practical Action International has opened a new office in Bolivia (Practical Action website, 2010).

The short-term duration of the CAB Regional Platform of Cleaner Production and the splitting off of their affiliated organizations after the funding of the German Technical Cooperation ended, show the dependency of actors from international cooperation agencies to keep the platform operational. Currently, cleaner production is not a priority for the German Technical Cooperation. The German Technical Cooperation has moved to other priorities such as democracy, rural development and water (GTZ website, 2010). Furthermore, the German Appropriate Technology Exchange (GATE) is not operational anymore. Moreover, actors working on appropriate and cleaner technology for small scale enterprises in Peru are disconnected and affiliated to several platforms and networks simultaneously. Self-interests among actors limit further cooperation (Sanchez, 2009, interview). At global level, the leadership of Practical Action is not disputed but the issue of appropriate technology is highly dispersed in several networks. For instance, the World Bank, the Inter-American development bank and Village Earth also promote appropriate technology in terms of renewable energy, information and communication technology for rural development and for rural small scale producers in developing countries. However, they do not work in cooperation with Practical Action International (IDB website, 2011; Practical Action website, 2010; Village Earth website, 2010; World Bank website, 2011). Additionally, the emergence of new networks of sustainable production, the formation of new platforms, and the lack of cohesion among actors contribute to dispersing of actors in the appropriate technology network.

The appropriate technology network show asymmetrical, horizontal and participatory relationships. The relationship of key actors with donor agencies is asymmetrical but their relationships with SMEs for transferring technology are of a more participatory nature.

7.2.3 Cleaner technology network

Key actors: the National Council of Science and Technology

The National Council of Science and Technology (Consejo Nacional de Ciencia y Tecnología, CONCYTEC in Spanish acronym) is the Peruvian governmental agency that regulates, coordinates and promotes science, technology and innovation oriented to enhance competitiveness and human development (CONCYTEC website, 2011).

The CONCYTEC's Cleaner Technology Programme promoted cleaner technologies in SMEs in Peru. This programme focused on knowledge generation and technology transfer, social appropriation of clean technologies, transformation processes, added value, bio-industry, natural products, fair trade and poverty fighting (CONCYTEC website, 2010; Sustainable Alternatives website, 2009+ Oliveros, 2006, interview). The programme was closed in 2005 due to lack of funding.

From 2006 onwards the support of CONCYTEC to SMEs was dispersed over several other programmes, including aquaculture, ICT, agriculture and agri-industry, medicinal plants, biotechnology, materials, vulnerability and the social dimension of science. Currently, CONCYTEC does not directly implement projects of technological transfer towards SMEs anymore. CONCYTEC promotes sustainable SMEs only indirectly, by establishing national research planning and guidelines, and providing funding for research (Huerta, 2011, interview). Funding for research on science and technology is operationalized in four schemes: projects of technological transfer (PROTEC), projects of innovation for competitiveness (PROCOM) and projects of scientific, technological research (PROCYT) and special projects (Paradigmas, 2010). Recently, CONCYTEC has also established the INNOTEC platform that promotes technological innovations that might potentially be applied to sustainable SME development (INNOTEC website, 2010).

Key actors: Technological Transfer Platform

The Technological Transfer Network (TTN) is a platform of five countries (Brazil, India, Nicaragua, Peru and Tanzania) that aims to exchange information on cleaner technologies and to enhance the transfer of cleaner and appropriate technologies for SMEs among developing countries. The platform has an online best practices database on cleaner and appropriate technologies called 'Sustainable Alternatives Network (SANET)'. The database has collected 1,473 case studies, 677 experts, 316 planning tools and 79 finance sources worldwide (Sustainable Alternatives website, 2010). The database is intended to be a resource for NGOs, universities, research centers and organizations that work directly with SMEs, especially rural enterprises (Sanchez, 2009, interview). Using the database, in 2005 UNEP offered a pilot online learning course on industrial energy efficiency (Sustainable Alternatives website, 2010).

The TTN Andean branch (called 'red Andina para la transferencia de tecnologias limpias', in Spanish) and the TTN Peruvian sub-branch focus on the implementation of regional multilateral agreements about Greenhouse Gases (GHGs) and Persistent Organic Pollutants (POPs). These mandates were discussed on the 4th meeting of the sub-regional committee of SMEs of the Andean Community (CAN) and the 14th meeting of the Andean Council of Science and Technology (CACYT) on November 2004 (CAN website, 2010). The Andean branch of TTN prioritizes the following sectors: mining, textile, garment, agri-industry, forestry and energy (CONCYTEC website, 2010). The TTN Andean branch and the TTN Peruvian sub-branch were hosted on the CONCYTEC webpage (www.concytec.gob.pe), but they are currently (2012) no available anymore.

Network relationships: cooperation

CONCYTEC cooperates with NGOs, universities, governmental agencies and international cooperation agencies to promote cleaner technologies in SMEs. CONCYTEC has played an active role in mobilizing and linking up local, national and regional actors to transfer technology to small scale industries and small scale producers.

CONCYTEC had an active participation in the implementation of the Technological Transfer Network (TTN) during the period 2004-2005. In March 2005 The Andean branch of the TTN platform held a meeting in Lima with UNEP to request additional financial support for the operational phase of the platform, this time including also Venezuela, Colombia and Ecuador, in addition to the initial countries (Sustainable Alternatives website, 2009+ Oliveros, 2006, interview). However, UNEP did not commit additional financial support to continue this project. CONCYTEC also collaborated in the implementation of the Horizontal Cooperation Programme on Cleaner Technology and Renewable Energy ('Programa de Cooperación Horizontal en Tecnologías Limpias y Energía Renovable', in Spanish) for 2005-2006. The Programme aimed to promote the adoption of cleaner technology and the use of renewable energy by micro and small enterprises in such a way that they can be prepared to apply for certifications (ISO 9000, ISO 14000 and SA 8000). The programme prioritized agri-industries (e.g. tanneries, dairies) and waste and sewage treatment plants. The programme was implemented in 9 Latin American countries (Argentina, Ecuador, El Salvador, México, Panamá, Paraguay, Perú, Trinidad y Tobago, y Surinam) (CONCYTEC website, 2010).

Till 2005, CONCYTEC had intensively collaborated with ITACAB, Practical Action Peru, Rainforest, Procabra and Energy, Development and Life (EDEVI) to support SMEs. These NGOs helped CONCYTEC to identify successful SMEs and best practices in SMEs. CONCYTEC had also ties with several national governmental agencies to reach particular SME sectors, especially agri-industries and manufacturing. CONCYTEC also established ties with domestic business associations (e.g. ADEX), banks (e.g. Peru Credito Bank), universities and technical schools (e.g. the National School for Industrial Training). At regional level, CONCYTEC cooperates with their pair institutions of science and technology of Brazil and Nicaragua at the TTN platform and with the aforementioned 9 Latin American countries participating in the Horizontal Cooperation Programme on Cleaner Technology and Renewable Energy (CONCYTEC website, 2009).

During 2004-2006 CONCYTEC strengthened collaboration with several international cooperation agencies promoting the transfer of cleaner technologies. The Organization of American States (OAS), the United Nations Environmental Programme (UNEP) and the Global Environmental Facility (GEF) are a few examples. They provided to CONCYTEC financial support and expertise (CONCYTEC website, 2009; Oliveros, 2006, interview). While UNEP, with funding support of GEF, established TTN and SANET, OAS established the Horizontal Cooperation Programme on Cleaner Technology and Renewable Energy in 2005.

With the collaboration of UNEP, in 2009 CONCYTEC and the Ministry of Environment launched the 2010–2021 national agenda for scientific research in climate change. The agenda calls for research on technology (modern, appropriate and traditional) to adapt to climate change (MINAM, 2010+ Huerta, 2011, interview).

Currently, CONCYTEC keeps ties with most of the aforementioned organizations but no collaborative work on cleaner technology has been identified. Although CONCYTEC works closer to ITACAB, the Ministry of Environment, the National Agrarian University and the National University of Engineering, it does not have connection with most NGOs and platforms of the appropriate technology network or the eco-efficiency network. Currently CONCYTEC focuses on providing funding for research projects at the national level, including for SMEs development (Huerta, 2011, interview).

Network relationships: power

This section analyzes the power relationship and distribution in the cleaner technology network, focusing on key organizations identified in Peru and their international connections. CONCYTEC is the national node of the cleaner technology network. TTN and OAS are the regional nodes and UNEP is the global node.

CONCYTEC has been the coordinator of the TTN Andean region during the pilot phase (till 2005) and the coordinator of the aforementioned OAS project till 2006. During this time, CONCYTEC brought together technological research centers and NGOs to implement demonstrative projects on cleaner technology, renewable energy and energy efficiency within small scale enterprises in Peru (CONCYTEC website, 2010). TTN is currently not as active in terms of coordination and cooperation as it was during the period 2004-2005. Most affiliated organizations joined TTN to gain legitimacy in public eyes and to explore opportunities of funding for their own projects. OAS promotes cleaner production and renewable energy in Latin America for SMEs in cooperation with UNEP, World Bank, the German Technical Cooperation and the Commission for Latin America and the Caribbean (OAS, 2004; OAS, website, 2010). In cooperation with OAS, UNEP designs projects and provides funding and guidance to governmental agencies, including CONCYTEC in Peru, for improving technology in SMEs worldwide. For instance, the implementation of the aforementioned projects by CONCYTEC has been done in close coordination with UNEP and OAS. By doing so, UNEP and OAS intended to enhance capabilities of governmental institutions on science and technology. However, expertise and access to funding from donors were controlled by UNEP and OAS, not by CONCYTEC. The position of actors (e.g. in being key partners and receivers of funding) depended from UNEP and OAS. SMEs were not part of such decision making processes on the promotion of cleaner technologies.

7.2.4 Technological innovation network

Key actors: Centers of Technological Innovation

Centers of Technological Innovation (CITEs) are private and public centers specialized in providing technological support, training and information for SMEs to increase their competitiveness and productiveness (PRODUCE website, 2011). CITEs are attached institutionally to the Peruvian Ministry of Production. In total there are 15 CITEs, each one specialized in a particular type of products, including, leather and shoemaking, wood and furniture, wine and horticulture, tropical fruits and medicinal plants, garment, agro-industry, textile, logistic and tracing, software and forest wood. The coordination of CITEs is centralized in a central office, labeled OTCIT in Spanish acronym. OTCIT is directly falling under the Peruvian Ministry of Production.

Network relationships: cooperation

The building up of cooperation with international cooperation agencies (e.g. IDB-FOMIN, the German Technical Cooperation and European Union), national lending institutions (e.g. COFIDE), governmental agencies, universities, companies, technical schools, NGOs and SMEs associations has been a priority for the OTCIT to fulfill its aims. OTCIT also has ties with the National Confederation of Private Business Associations (CONFIEP in Spanish Acronym), the Pontifical Catholic University of Peru and the NGO GS1Peru.

Most CITEs are NGOs. They provide services to SMEs in quality control of raw materials and final products, design and product development, production technology and environmental management. For instance, CITE Agroindustrial CEPRORUI and CITE Confecciones are led by El Taller and CITE Agroindustrial Piura includes CEPICAFE and producer associations of Piura (CITEs website, 2011).

CITEs have ties with the Ibero-American Programme for Science, Technology and Development (CYTED in Spanish acronym) and the Management, Evaluation, Monitoring of Results in Science, Technology and Innovation to Improve Impact in Development (RESIRDES), two regional platforms promoting science and technology. CYTED is a platform of national governmental agencies of science and technology of 21 countries of Latin American, plus Spain and Portugal. It facilitates knowledge exchange and finances research projects on science, technology and innovation. RESIRDES is a platform to exchange experiences on technology and innovation for rural development at regional level in Latin America. RESIRDES affiliates ministries and national governmental agencies on science and technology and universities. One of the priorities of the platforms is to promote research on modern technologies for industrial development and cleaner technologies (CYTED website, 2010; PRODUCE website, 2011).

The International cooperation agency of Spain and the German Development Service (DED) are the main agencies providing funding to CITEs. AECI specially supports CITEccal (Leather and Footwear Innovation Center), CITEmadera (Wood and furniture) and CITEvid (grapes & wine) since their launching in 1999. The German Development Service also provides financial support to CITEmadera (Carazo, 2006, interview; Peru Innova No55, 2010; Peru Innova No56, 2010; PRODUCE website, 2011). Other funding sources are FINCYT and FIDECOM. FINCYT, a national funding scheme for science and technology, has been established by the national government and the Inter-American Development Bank in 2007. FIDECOM is a national fund to promote research in production competitiveness and innovation in SMEs. FINCYT and FIDECOM are administered by CONCYTEC.

A set of organizations including NESST and CONDESAN have launched a new project that promotes technology and innovation for development in Peru: RAMP Peru. The RAMP Peru project promoters cooperate with CONCYTEC, FINCYT and RESIRDES, but no cooperation was found with CITEs (NESST Peru website, 2012).

Network relationships: power

The national node of the network is OTCIT. RESIRDES is the regional node. At global level, the nodes are the Spanish Agency for International Cooperation and the German Development Service.

Although each CITE has independency to work, OTCIT coordinates with the governmental agencies, fellow international organizations on technological innovation and funding agencies. OTCIT has close ties with RESIRDES. The Spanish Agency for International Cooperation and the German Development Service are the main supporters of CITEs. Resources in terms of funding and expertise are mainly hold by the Spanish Agency for International Cooperation, the German Development Service, the Ibero-American Programme for Science, Technology and Development, the German Technical Cooperation, the European Union and the Inter-American Development Bank. At the national level, CONFIEP is a key actor. SMEs participate in CITEs. At the national level, CITEs provide services for SMEs that are formally registered and have the capacity to scale up production (PRODUCE website, 2011). Major disputes were not identified.

7.2.5 Urban cleaner production network

Key actors: IPES Promoting Sustainable Development

IPES Promoting Sustainable Development (IPES Promoción del Desarrollo Sostenible in Spanish) is a Peruvian NGO founded in 1984 that aims to improve the living conditions of less privileged people and to build sustainable cities. The core issue of expertise of IPES is waste management at local level. In cooperation with national partners, IPES works in several Latin America countries, including Colombia, Cuba, Brazil and Argentina. IPES has implemented several projects promoting cleaner technologies and good practices in Peruvian urban small scale enterprises, to increase their competitiveness.

Network relationships: cooperation

With the financial support of the Dutch Catholic Organisation for Relief and Development Aid (CORDAID, English acronym), during 2001-2004 IPES implemented the project 'good environmental practices in small scale enterprises of the Villa El Salvador's Industrial park'. During 2003-2005 IPES implemented a project with financial support of the Inter-American Development Bank to improve productiveness and environmental performance in small scale shoemakers and tanneries in the city of Trujillo, in the northern part of Peru. The financial contribution of the Inter-American Development Bank was 66% and that of IPES was 34%. With the financial support of the Fighting against Poverty in Metropolitan Lima Programme (PROPOLI), during 2004 and 2005 IPES implemented two projects to improve production processes, working conditions and risk prevention within small scale carpenters and metal workshops in Villa El Salvador. PROPOLI has been implemented in Peru by the Ministry of Woman and Social Development (Ministerio de la Mujer y Desarrollo Social, MIMDES in Spanish acronym) with the financial support of the European Union. The cooperation of small scale enterprise associations and municipalities has been valuable to implement PROPOLI. Since 2006, however, IPES has stopped implementing projects on cleaner production, and has returned to its core business: consultancies on waste management at the local level. This service is provided for municipalities, national governmental agencies and international governmental agencies. At the international level, IPES has established cooperation ties with the International Development Research Center (IDRC), the Swiss Federal Laboratories for Materials Testing and Technology (EMPA), the organization for the Ecology and Development (ECODES) and the Dutch consultancy WASTE to explore opportunities to establish micro scale enterprises to recycle electronic waste (IPES website, 2012).

Network relationships: power

The main national actor in the urban cleaner production network is IPES. At regional level, the central node is the Inter-American Development Bank. At global level, the main nodes are the European Union, the Dutch Catholic Organisation for Relief and Development Aid and the State Secretariat for Economic Affairs of the Swiss Federation. At national level, key allied organizations of IPES are municipalities and the Ministry of Environment of Peru. IPES has no links with the Latin American NCPCs or RESIRDES. Most projects on cleaner production implemented by IPES were directly financed by international cooperation agencies. This direct cooperation with donors gives IPES a good position in the network; however, its work is quite isolated from national platforms and actors involved in sustainable production. Major disputes were not identified in the network.

7.2.6 Networks: patterns, challenges and trends

In this section the main patterns, challenges and trends of the sustainable production networks are analyzed. The main issues to be analyzed are cooperation, power relationships, the roles of NGOs, key challenges and significant trends.

Cooperation

Cooperation of actors in the sustainable production networks has the following is characterized by contrasts. Three main charactersitics are identified in the networks: (1) overlap in the promotion of sustainable production; (2) dependency on funding and expertise; and (3) distrust and low effectiveness.

   First, the five sustainable production networks overlap in their efforts to promote sustainable production into SMEs. NGOs and (inter)governmental agencies promoting sustainable production are dispersed and disconnected. Key actors at national level have scarce ties, and little joint cooperation. CONCYTEC has ties with CITEs but no ties were identified with Practical Action Peru, CER or IPES; similarly ITACAB has ties with Practical Action and CITEs but not with CONCYTEC, and neither with CER as far as cleaner production is concerned. IPES does neither have ties with CITEs nor with CER. Moreover, the connections of key NGOs and key (inter) governmental agencies with SMEs are also scarce and on a temporal basis. Moreover, actors over time can abandon a platform, join a new platform or be affiliated in several platforms simultaneously. This strong mobility of actors in the networks, however, does not match with the scarce cooperation. For instance, CONCYTEC and Practical Action Peru were part of both the TTN platform and the CAB Regional Platform of Cleaner Production but they did not build ties mutually, nor with the platforms. Furthermore, key actors of the sustainable production networks have scarce cooperation with most NGOs working on the sustainability of SMEs. For instance, ITACAB and Practical Action promote appropriate technologies for sustainable production but no cooperation was found with actors of other platforms of the appropriate technology network, and also not with other sustainable production networks. Additionally, no ties have been found with RAE Peru, RAAA, ANPE and COPEME, which also works with small scale enterprises and small scale producers (see Chapter 5). Global actors also overlap in their efforts, although they have more cooperation and exchange of resources. Most important international cooperation agencies involved in funding and policy guidance for key actors in the sustainable production networks have their own intervention plan and are connected to a particular set of national NGOs and (inter) governmental agencies. The most evident example of this overlap is UNIDO and UNEP.

Second, operations of actors and platforms in the sustainable production networks depend strongly of funding and expertise to operate successfully. Sustainable production platforms do not have committed 'self-mobilizing' actors. Once the project ends, actors and platforms become 'latent' waiting for the next funded project. For instance, CONCYTEC played a relevant role in linking up actors and implementing 'pilot' projects of cleaner production in SMEs under the umbrella of the TTN platform, only during the period that funding and expertise from UNEP and the Organization of American States were available. According to Huerta (2011, interview) 'the bottleneck of the application of the National System of science, technology and technological innovation (CINACYT) for transferring technology to SMEs has to do with the lack of financial budget and highly qualified researchers'. Moreover, the agenda of priorities of actors at national level depends on the intervention agenda of international cooperation agencies. Additionally, key actors of the sustainable production networks compete for funding to operate projects of international cooperation agencies. The failure of the TTN platform and the CAB Regional Platform of Cleaner Production has shown that platforms in the sustainable production networks cannot survive independently of funding. The TTN platform and the CAB Regional Platform of Cleaner Production switched to 'sleeping mode' after the financial support ended in 2006. What has been left operating of the CAB Regional Platform of Cleaner Production is the online database of cleaner and appropriate technologies for small scale enterprises which is currently being used as resource for new projects of ITACAB. While the CAB Regional Platform of Cleaner Production was operative it acted as 'glue' to bring together affiliated organizations. The main actors, platforms and promoters of the networks are summarized in Table 7.4.

Third, the sustainable production networks are characterized by mistrust and low effectiveness. SMEs and national key actors do not keep long lasting cooperation because SMEs do trust neither NGOs nor (inter)governmental agencies. The trial-error intervention of NGOs has created a bad image of NGOs among SMEs (Oliveros, 2006, interview; Sanchez, 2009, interview). National actors also reach only a small amount of SMEs. While some SMEs, mainly the exporting and servicing SMEs, are interested in cleaner production because of pressure of international markets, most SMEs do not realize the value of cleaner production for their business (Su, 2004+ Espinosa, 2008, interview). In the words of Sanchez (2009, interview): 'technology is available but the problem is the social and political dimension of technology. (...) several small scale enterprises and

Table 7.4. Networks promoting sustainable production for SMEs in Peru.

Networks	National actors	Regional platforms	Regional/global actors
Eco-efficiency	CER	Latin American NCPCs	UNIDO, SECO
Appropriate technology	ITACAB, Practical Action Peru	CAB Platform of Cleaner Production	GTZ, Practical Action International
Cleaner technology	CONCYTEC	TTN	UNEP, OAS
Technological innovation	CITEs, OTCIT	RESIRDES	AECI, DED
Urban cleaner production	IPES	-	IDB, EU, CORDAID

cooperatives in communities have failed due to individualism and scarce cooperation of people, for instance, the ECASH-MAYA project and the PROSAMA project'. The low capacity of NGOs to implement projects has also been questioned by international cooperation agencies (Kuriger, 2006, interview). Additionally, the emergence of new networks, the structuring of new platforms, the lack of managerial capabilities of key actors and the lack of economic incentives for SMEs to adopt sustainable production contribute to failures or stagnation in the adoption of sustainable production by SMEs in Peru.

Power relations

Power relations in the networks of sustainable production have two main characteristics: (1) the asymmetrical relationships of actors; and (2) the power of SMEs in refusing or accepting sustainable production.

First, power distribution is asymmetric in the networks and it is expressed in the division of roles in national, regional and global nodes (Figure 7.1). The Inter-American Development Bank, Practical Action International and the Dutch Catholic Organisation for Relief and Development Aid CORDAID have the capacity to control resources in terms of agenda setting, expertise and funding provision. National key actors also have the capacity to control resources but to a lesser extent and in relationship of mutual dependency with global key actors. National key actors have more connections and exchanges with international cooperation agencies than with SMEs. This is related to the fact that national key actors and platforms depend on funding and expertise from international cooperation agencies.

Second, the scarce motivation of SMEs to buy-in sustainable production puts them in a power position to delegitimize interventions of promoters of sustainable production. Promoters have good

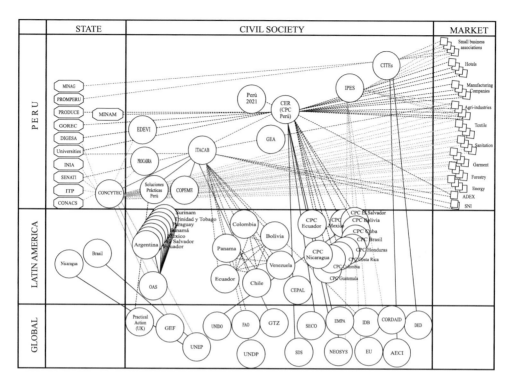

Figure 7.1. Map of the networks of sustainable production, including their key actors, platforms and connections.

will but lack the capacities to build lasting ties with SMEs through the platforms. SMEs have the capacity to refuse or accept sustainable production. This capacity is founded in the embeddedness of SMEs in the local social networks. SMEs with their owner, manager and employees have strong ties with social groups, families and communities where they belong to (Cici & Ranghieri, 2008). This contributes to the strength of SMEs in delivering goods and services for national and local markets in Peru and other countries of Latin America, despite of their constrains in terms of organization and business capacities (Corral *et al.*, 2005). Indeed, the embedding of SMEs in local social networks has contributed to the successful scaling-up of SMEs in Peru, including those connected to the cleaner production networks. So, both promoters and SMEs hold power in the sustainable production networks. While the power of promoters rests on the capacity to develop connections of wide scope, establish global agendas, give policy guidance and provide funding and other resources to govern the networks, the power of SMEs rests on their capacity to be anchored within the local social networks. While SMEs are skilled in building ties within their local network, most of them have little interest in expanding their connections to sustainable networks.

Roles of NGOs

Key actors in the sustainable production networks are involved in the promotion, adaptation, innovation and transfer of technology-based approaches, including eco-efficiency, renewable energies and low-cost technologies for rural and urban SMEs and micro scale enterprises. In promoting sustainable production, NGOs focus on particular type of SMEs. For instance, CER has more cooperation ties with exporting SMEs, CITEs have more ties with micro and small producers, Practical Action Peru prioritizes rural micro and small scale enterprises oriented to local markets and IPES prioritizes small scale enterprises in urban areas. While NGOs typically play central roles as national key actors in the eco-efficiency network, the appropriate technology network and the urban cleaner production network, (inter)governmental agencies play central roles in the cleaner technology network and the technological innovation network (Table 7.4). Typical NGOs in the sustainable production networks are technocratic NGOs working with SMEs in the rural and urban areas. (Inter)Governmental agencies performing as key actors at the national level appear to be less effective in promoting sustainable production than the technocratic NGOs.

Challenges

The main challenges to promote sustainable production practices in SMEs have to do with two factors: (1) the lack of conditions that enable SMEs to adopt sustainable production; and (2) the underestimation of the innovative power of SMEs.

    First, most SMEs operate according to a business-as-usual frame. It means that SMEs prefer to obtain low profit now rather than later, even if later they get more benefits. In the words of Espinosa (2008, interview) 'the outcome of eco-efficiency improvements is at the long-term but most SMEs want increase productiveness and return of investment at the short-term. SMEs invest in new technologies if the pay back time of the investment is short. Costly technologies require several years to recover'. SMEs prefer to invest in second-hand machinery. SMEs lack the culture of saving. Most SMEs are not interested in up-to-date technology, even if credit schemes for these technologies are available. Cleaner technologies are financially less appealing for SMEs, especially in countries like Peru with high interest rates and short time span to bring back the lending (CER, 2008). Additionally, it should be pointed out that credits for SMEs to upgrade technology reach a very limited number to SMEs due to the requirements of banks. Besides, although eco-efficiency is in line with business interests, most SMEs are not interested in it, and only do what they are obliged to by regulation. According to Alegre (2011, interview), director of CER, 'Peruvian business people have inconsistent thinking. They say the Peruvian economy is growing but what is the situation of education and environmental consciousness? I mean the overall business culture is very low, at least we are 20 years backwards in comparison to more advanced countries'. Moreover, the adoption of cleaner production by micro and small enterprises that produce for national and local markets is harder, especially if they are not interested (Bickel, 2006, interview). Furthermore, most SMEs are informal enterprises with limitations to access capital and technology, low awareness of environmental management and lack of a policy framework for cleaner production adoption (Espinosa, 2008, interview; Oliveros, 2006, interview; Vidal, 2006, interview).

Second, the involvement of SME in participating in designing policy interventions is missing in the five networks of sustainable production. Only in the appropriate technology network and the urban cleaner production network we saw participatory schemes of cooperation with small scale enterprises. Overall decision making about interventions is reserved for international cooperation agencies and national key actors. This lack of participation in the network structure hampers the innovative power of SMEs. SMEs are dispersed and fragmented in hundreds of small business associations. Moreover, usually medium-sized enterprises are only engaged with larger companies and small and micro scale enterprises are clustered in many tiny business organizations (Oliveros, 2006, interview; Sanchez, 2009, interview). SMEs do not have organizational structures to participate in networks' decision making.

Trends

The following three trends are detected in the networks of sustainable production: a further adoption of cleaner technologies by SMEs engaged to international markets as suppliers; stagnation in cleaner production for SMEs oriented to national and local markets; and further collaboration of actors at the global and regional level. SME beneficiaries of sustainable production networks include dairy industries, carpenters, metal workplaces, (organic) producers and craft-makers. They can be urban and rural small scale enterprises, including those oriented to local, national and international markets. Most of the textile, garment, agri-industry and forest-based enterprises oriented to global markets have the resources and the capabilities to change to sustainable production. They can apply for subsidies, green credits and grants to go for such a change. Exporting SMEs are the target of key actors in most networks, except the appropriate technology network and the urban cleaner production network. However, most other SMEs are difficult to engage in sustainable production without the use of incentives and subsidies. Incentives and subsidies are expected to encourage SMEs oriented to local and national markets to adopt sustainable technologies. However, these instruments cannot produce results without investigation of the motivations and enabling conditions of micro and small scale enterprises to adopt sustainable technologies. Finally, looking at the recent joint efforts of UNIDO and UNEP to establish the global platform RECP, it can be said that further join cooperation of regional and global key actors involved in the promotion of cleaner production is expected.

7.2.7 Conclusions

a. The eco-efficiency network, the appropriate technology network, the cleaner technology network, the technological innovation network and the urban cleaner production network are the main networks involved in the promotion of sustainable production for SMEs in Peru. Due to the lack of trust among actors, the five cleaner production networks show very limited cross ties and cooperation. Most national actors are not connected to each other and compete for scarce resources. As a result, key actors show strong duplication of efforts, and low efficiency and efficacy in promoting sustainable production. For instance, CER does not have ties with most NGOs working with SMEs. Practical Action Peru, ITACAB and IPES focus on small and micro industries but scarce cooperation was identified between the first two. IPES is isolated from the rest of national key actors. Similarly CER and CITEs prioritize medium and larger industries but no cooperation was identified with the rest of the key actors.

b. Platforms are established at national, regional and global level by international cooperation agencies to centralize coordination. Platforms are usually online portals and not spaces for collective 'self-mobilizing' of actors. TTN and SANET are online portals of value for networking and exchange, but hardly can nest self-organization and self-mobilization of NGOs and SMEs. If a platform has 'organic' life, it usually has short-term duration. SMEs have scarce participation in the platforms and in the decision making of agenda's of key network actors. Additionally, the building of links with the Peru's exporters association and other business associations has been a good starting point, but they affiliate only small amount of SMEs.

c. Global key actors have higher control of resources in the networks but the scarce motivation of SMEs to adopt sustainable production delegitimizes them. Expertise and funding are concentrated at the global nodes. National key actors show high dependency of funding and expertise to implement projects of sustainable production in SMEs. Top-down approaches towards SMEs are common in the networks to allocate technology, rather than more participatory approaches. Networks do not show committed SMEs promoting cleaner production hand-in-hand with NGOs. SMEs are dispersed, fragmented and most of them are not connected to the cleaner production networks. Networks fail in building lasting ties with SMEs. However, the study shows that SMEs engaged to international markets as suppliers have better chance to adopt sustainable production, compared to SMEs oriented to the local and national markets. SMEs reveal their power by either refusing or adopting sustainable production. Thus, power does not only operate in the networks as the capacity to allocate resources, but also as the will to accept or refuse sustainable production measures.

d. The types of NGOs involved in cleaner production networks are diverse. CER and CITEs are 'technocratic' NGOs that have expertise in process engineering, design and technology-based solutions. Practical Action Peru and IPES are technocratic NGOs with expertise in participation and production technology for SMEs. RAMPERU is also a new NGO promoting technology for development in Peru. CONCYTEC and ITACAB are '(inter)governmental' agencies that have expertise in promoting national policies and research on science and technology. Although CONCYTEC and ITACAB are not typical NGOs, they channel funding and operate projects of international cooperation agencies. Those properties of both (inter)governmental agencies make them cluster as a sort of government organized non-governmental organizations (GONGOs). Therefore, NGOs and Government Organized NGOs in the networks of sustainable production perform as intermediaries between 'promoters' of sustainable production and SMEs 'beneficiaries' by operating 'pilot' projects and providing advice to national policy makers, other NGOs, international cooperation agencies, national environmental authorities and SMEs on eco-efficiency, renewable energies and low-cost technology.

7.3 Discourse analysis of sustainable production

7.3.1 Introduction

Agents are basic elements of discourses and central actors of networks. Additionally, it should be kept in mind that sustainable production discourses do not match necessarily one-to-one with the networks of sustainable production described in the network analysis section.

Based on the Dryzek's (1997) definition of discourse, cleaner production can be seen as a diverse set of assumptions, judgments and contentions. Each discourse of sustainable production is a shared way to apprehending improvement on production efficiency encompassing plausible storylines or accounts about production and the market by agents including NGOs, SMEs and micro-enterprises, business associations, governmental agencies and international cooperation agencies. The discourse analysis of cleaner production focuses on views about production, the market and environmental sustainability, and the role of key agents involved in making production of SMEs sustainable.

The discourses of sustainable production are analyzed following the methodology of Dryzek (1997), as described in depth in Chapter 3. Four elements are analyzed to construct storylines: (1) basic entities acknowledged; (2) assumptions about natural relationships; (3) agents and their motives; and (4) key metaphors and other rhetorical devices.

Discourses on sustainable production

Two discourses are identified: the discourse of 'cleaner production' and the discourse of 'appropriate technology' (Table 7.5). Both discourses have communalities and differences. The communalities are that they endorse the central role of technology to bring about sustainable SMEs. Sustainable SMEs are understood in both discourses as SMEs with higher revenues, higher productivity and more eco-efficient production processes. Also, both discourses identify the same agents including 'technocratic' NGOs, (inter)governmental agencies (GONGOs), international cooperation agencies and frontrunning SMEs. The difference between the discourses is in the focus. While the cleaner production discourse focuses on up-to-date modern technology to increase production efficiency of SMEs, the appropriate technology discourse focuses on small scale technologies suitable to the needs of small and micro scale enterprises.

The cleaner production discourse highlights the role of modern technology in increasing efficiency, productivity and environmental performance of SMEs. The discourse focuses on SMEs supplying international markets or with capacities to do so. The main motivation of the discourse proponents is to encourage SMEs towards sustainability by improving eco-efficiency in production and connecting SMEs with international markets. By doing so, key agents aim to stimulate environmental friendly practices in SMEs without stopping business growth.

The appropriate technology discourse highlights the role of tailor-made technology to increase productivity and environmental performance of SMEs, particularly micro and small scale enterprises. Low capital, and small scale technology suitable in the local social and cultural setting are central in the discourse. The discourse highlights the use of renewable energy, development of local and national markets and poverty fighting, regardless of enterprise size, economic

Table 7.5. Overview of discourses on sustainable production.

Discourse	Cleaner production	Appropriate technology
Storyline	Cleaner production is seen as business strategy to make SME production more efficient. Allocate the best up-to-date modern technology as possible is central to reduce environmental impacts and increase competitiveness. The discourse focuses on SMEs that are well established at the national market and with capacities to reach international markets.	Appropriate technology is seen as tailor-made technology adjusted to the needs of SMEs, particularly micro and small scale enterprises. Low capital, small scale and suitable technology for the local social, economic and cultural setting are central in the discourse. The discourse highlights the use of renewable energy, development of local markets and poverty fighting.
Type of NGOs	GONGOs Technocratic urban NGOs	GONGOs Technocratic rural NGOs Technocratic urban NGOs
Subscriber	CER, CONCYTEC, CITEs, CESEM	ITACAB, Practical Action Peru, IPES

activity or geographical location. The main motivation of the discourse proponents is to make the poor people's micro and small scale enterprises and their communities sustainable, based on technological self-reliance. Key agents establish technological solutions in a participatory way to improve SMEs workers' and communities' life conditions.

The cleaner production discourse and the appropriate technology discourse come from different traditions and schools of thought. The cleaner production discourse has its roots in the environmental concerns of industrial development such as the control of industrial pollution and industrial waste minimization. The term cleaner production was coined by the United Nations Environment Programme in 1989 as a strategy for industries to contribute to sustainable development. The appropriate technology discourse has come up from the ideas of Gandi's philosophy and Schumacher's thoughts. This discourse endorses a more autonomous development of rural communities regarding energy provision, building construction, water supply and treatment, sanitation and food production, and information and communication technologies.

In the next sections the discourse of cleaner production and the discourse of appropriate technology are developed in detail.

7.3.2 Discourse of cleaner production

Enhancing competitiveness of SMEs through cleaner production!

Cleaner production is seen as business strategy to minimize loss of raw materials, improve energy efficiency, increase water efficiency and waste minimization, and better house-keeping. The discourse claims that the reduction of environmental impacts and the increase of productivity and competitiveness of SMEs can be achieved by the provision of modern up-to-date technologies. An overview of the discourse of cleaner production is summarized in Table 7.6.

Basic entities acknowledged

The liberal market is acknowledged. The liberal market is understood as opportunity for business growth, technological upgrading, innovation and increased business competitiveness (Alegre, 2009; TTN-Red Andina, 2005). The liberal market enhances connections of SMEs, especially, with national and international company buyers (Alegre, 2007; CER website, 2011). The connection is understood in terms of exporting locally produced goods and importing technological knowledge and devices to improve the efficiency of production processes. The connection of SMEs to highly profitable national and international markets is seen as requirement to become sustainable (CESEM

Table 7.6. Discourse analysis of cleaner production.

Elements	Cleaner production
Basic entities acknowledged	Liberal market
	Eco-efficient processes and technology
	Environment
	Others: national NGOs, government, international cooperation agencies, intergovernmental agencies and financial institutions
Assumptions about natural in relationships	Hierarchy
	Efficiency
	Nature is understood in mechanistic terms. Environmental improvement will be a consequence of the implementation of environmentally sound technology
Agents and their motives	National NGOs: service providers to SMEs
	International cooperation agencies: Provide policy guidance, expertise and funding to increase SMEs competitiveness
	Exporting SMEs: international markets, economic growth
	National government: economic growth and competitiveness
Key metaphors and other rhetorical devices	Rhetorical devices: 'technological transfer', 'eco-efficiency', 'competitiveness'

website, 2010; Flores, 2004; TTN-Red Andina, 2005). In this sense, the cleaner production discourse endorses free trade agreements and the engagement of the national economy with the world market.

Cleaner technology and eco-efficient processes are acknowledged. Technological diffusion, transfer and upgrading are considered central to increase economic competitiveness and environmental improvement of SMEs (Flores, 2004; PNUMA-CONCYTEC, 2004; TTN-Red Andina, 2005). Eco-efficient production processes avoid costly solutions of pollution control such as waste management and air pollutions. By doing so, companies reduce their cost of production and the quality of the end-product (Alegre & Marthaler, 2010). Improving eco-efficiency of SMEs does not necessarily mean sophisticated changes, but can also go via easy changes in production patterns and working styles. According to their proponents, cleaner technology and eco-efficient processes improve production capacity and enhance more standardized production of SMEs to satisfy the demands of highly profitable markets (Alegre, 2007; CER website, 2011; IPES website, 2012).

The environment is acknowledged. Environmental protection is central in the discourse. The discourse embraces the idea that the only option to prevent the environmental burden created by modern industrial activities is to move to more environmental sound technologies and more efficient use of raw materials, water and energy (Alegre, 2009; Huerta, 2011, interview). Environmental improvement is seen in terms of environmentally sound technological upgrading, tuning production processes, change and recycling of materials, and re-designing of products (Alegre, 2007). Proponents highlight that cleaner production is an environmental protection strategy from an economic point of view. According to the cleaner production discourse, environmental protection is feasible in SMEs as long as it enhances economic competitiveness (IPES website, 2012). Then, environmental benefits are a positive side-effect. This has been established by several discourse proponents. According to Alegre (2011, interview), director of CER, 'the promotion of eco-efficiency in companies contributes to their competitiveness'. In these terms, cleaner production is seen as the operationalization of sustainable development in the production domain.

Other basic entities recognized in the discourse are network of NGOs, national government, international cooperation agencies, inter-governmental agencies and financial institutions.

Assumptions about natural relationships

One assumption in the cleaner production discourse is on hierarchy, based on the adoption of cutting-edge modern technology. Technology establishes hierarchical positions among enterprises depending on their capacity to meet the demands of global markets (CONCYTEC, 2006; Flores, 2004). Transnational companies, larger companies and SMEs holding modern technology and sophisticated production process are situated at the top. Contrarily, national industries with old fashioned technology and inefficient production processes are situated at the bottom. Global and local powerful market actors set the requirements for supply of goods and services. In these terms, the only option to increase environmental performance and make more competitive SMEs is to provision the best environmental sound technology.

Efficiency is understood in the cleaner production discourse as the higher physical and economic optimization of industrial production processes. Efficiency is at the core of cleaner production (Espinoza, 2008, interview). In the eyes of the discourse proponents, the efficient use of raw materials, water and energy determine better productiveness. This matching of industrial efficiency with the conservation of natural resources is referred by the proponents of cleaner production discourse as eco-efficiency.

Regarding the relationship with nature, it is understood in terms of machinery, engineering processes and economic rationality. Protecting nature is seen in terms of a more efficient use of natural resources and less generation of waste in production processes. In the cleaner production discourse nature is reduced as input of production.

Agents and their motives

Agency is granted to national NGOs. NGOs operating at the national level raise awareness in local industries and in their associations, provide training for technicians, carry out in-plant assessments and promote investment for technology upgrading of SMEs. In cooperation with international cooperation agencies, national NGOs implement demonstration projects within SMEs; for instance, to prevent industrial pollution and diminish Green House Gas (GHG) emissions (Cortijo, 2006, interview; Espinoza, 2008, interview). Other NGOs are incorporating the territorial dimension in promoting cleaner production. For instance, CER is implementing eco-industrial parks in several parts of Peru. National NGOs intend to fill the technological gap of SMEs by linking cleaner production with social responsibility standards, environmental management systems (EMSs) and initiatives of Clean Development Mechanisms (CDMs).

Agency is also granted to international cooperation agencies, including international NGOs, intergovernmental agencies, international financial institutions and governmental agencies of developed countries. International cooperation agencies channel resources to national NGOs, governmental agencies and business associations of developing countries to support SMEs. For instance, most research on science and technology in Peru is funded by international cooperation agencies, not by the national government (CONCYTEC, 2009). Additionally, international cooperation agencies establish the agenda of priorities in the networks of sustainable production. Most international cooperation agencies prioritize SMEs that are tied to exporting value chains, not the ones oriented to local markets only (COSUDE/SECO website, 2011; IDB website, 2011; UNIDO website, 2011; UNEP website, 2011).

International cooperation agencies such as the United Nations Industrial Development organization (UNIDO) and the United Nations Environmental Programme (UNEP) are central promoters of cleaner production worldwide. For instance, UNIDO facilitates cooperation among National Cleaner Production Centres (NCPCs), between NCPCs and the national governments and between NCPCs and SMEs. International cooperation agencies also establish policy guidance and develop service packages to better respond to the needs of SMEs. UNIDO and UNEP implement multilateral environmental agreements such as the Kyoto Protocol and its Clean Development Mechanism, and support governments to develop the national environmental legislations on cleaner production (UNEP Institutional website, 2010; UNIDO Institutional website, 2010). International cooperation agencies and market actors pressure and condition developing countries for industrial change, for instance, textile industries to improve their environmental performance. The motivation of international cooperation agencies in promoting cleaner production is to increases the competitive position of local SMEs in the globalizing economy, meet environmental compliance and create a market of cleaner production services for SMEs (Cortijo, 2006, interview).

Agency is limited to competitive SMEs, not relevant for all SMEs. The discourse recognizes as agents SMEs (most of them medium-sized enterprises) that are mainly suppliers to domestic or exporting larger companies. SMEs subscribing the discourse are, among others, manufacturing, agri-industries, textile-industries, garment workshops, forest based industries, shoemakers and tanneries (Alegre, 2007; Febres, 2004; Hechaverria, 2004; IPES website, 2012; Sibille, 2004). SMEs can be individual SMEs or SMEs organized in industrial parks. Competitive SMEs have gained a status in the eyes of the discourse proponents. They are referred by technocratic NGOs and GONGOs subscribing this discourse as 'clients', not so much as 'beneficiaries'. The motivation of SMEs and their business associations to implement cleaner production measures are mainly economic, such as increase profit, increase competitiveness and reach foreign markets (Alegre, 2007). SMEs subscribing to the discourse have the willingness and the capacity to pay the costs of making their production process cleaner (Alegre, 2011, interview). However, most companies, including SMEs and microenterprises, do not show much interest in adopting cleaner production measures (Bickel, 2006, interview). As Alegre stated (2011, interview) 'at first glance cleaner production is catchy, for instance saving money by implementing eco-efficiency; but in practice companies do not buy-in it easily. Companies at best only do what they are obliged by the national environmental regulation. Most companies have short-term thinking'. Moreover, discourse proponents consider most small scale enterprises and micro enterprises as enterprises with a lack of capacity to handle technology and implement eco-efficiency measures in production processes (Sanchez, 2009, interview).

National government has agency but it shows limited leadership in making local SME production cleaner and more sustainable. National government has short-term views of SME development and shortcomings in implementing environmental policies in the national production domain. The national environmental policy is less focused on law enforcement and more on the promotion of voluntary strategies. Central national government does not stimulate conservation of natural resources in SMEs, less the promotion of environmental standards (CONCYTEC, 2009). Most policy makers, especially at the central government, see environmental protection as an obstacle for investments, economic growth and job generation. Economy and ecology do not have the same weight in policies towards SME development. For instance, the national government puts lower national environmental standards for SMEs in order to generate more jobs. However, the discourse proponents recognize the efforts of the Ministry of Environment of Peru in developing the national environmental policy framework, especially regarding cleaner production. According to Marcos Alegre (2011, interview) the Ministry of Environment of Peru has shown commitment with eco-efficiency thanks to the influence of CER and international cooperation agencies. For instance, it has played a decisive role in making eco-efficiency compulsory for public agencies. Discourse subscribers claim the need of economic incentives for SMEs to improve their environmental performance, but this is not easy in the economic model applied in Peru. As Alegre stated (2011, interview) 'subsidy schemes depends of the Ministry of Economy and Finance of Peru (MEF, in Spanish acronym). This kind of measures is very hard to adopt under free market conditions'.

Key metaphors and their rhetorical devices

The cleaner production discourse includes key rhetorical devices. The rhetorical devices identified are 'technological transfer', 'eco-efficiency' and 'competitiveness'. Technological transfer refers to the sharing of environmentally sound technologies, knowledge, manufacturing methods and expertise, usually from developed countries to developing countries. The transfer aims to create a service market for SMEs in developing countries. Eco-efficiency is located at the core of the cleaner production discourse. Eco-efficiency is related to the properties of technology: to minimize ecological damage while maximizing efficiency of production process. Finally, competitiveness cannot be missed in any claim of the cleaner production discourse (CER website, 2011). Competitiveness refers to the ability of SMEs to sell and supply goods and services profitably on national and international markets. Competitiveness in the discourse is related with economic growth, productivity and export capacity of SMEs.

7.3.3 Discourse of appropriate technology

Appropriate technology to fight poverty!

Appropriate technology is seen as technology suitable to the specific needs of SMEs, particularly micro and small scale enterprises. The discourse stresses the idea that small scale, low capital technology and local knowledge together cope better with the sustainability challenges of poor people's enterprises. An overview of the discourse of appropriate technology is summarized in Table 7.7.

Basic entities acknowledged

The liberal market is acknowledged. The liberal market is seen as a given powerful force that brings threats and opportunities for SMEs. It becomes an opportunity as long as SMEs adapt the available technologies to their needs in order to create a market for their products (ITDG, 2009). The liberal market becomes a threat as long as SMEs apply the available technologies without evaluating their sustainability. According to Carlos De La Torre, project coordinator of Practical Action Peru, the access to markets for small scale and micro enterprises is central (Soluciones Practicas website, 2011).

Small and low capital technologies are acknowledged. Technology is understood as low cost gadgets, easy to use, maintain and handle. Technology can also be sophisticated but it has to be adapted to local people in terms of their social setting, cultural background and economic conditions (ITDG, 2009; Sanchez, 2008). The transfer or innovation of technology is done with close participation of micro and small scale enterprises (ITACAB website, 2011). Technology is seen in the discourse as a mean to connect small scale producers to local and national markets rather than the international market (Sanchez, 2009, interview). Despite the fact that appropriate technology is based on science, it deviates from industrial technology which is highly sophisticated and standardized, large scale, and capital and energy intensive. Appropriate technology also differs from traditional technologies that are usually seen as inefficient and unproductive, aimed for

Table 7.7. Discourse analysis of appropriate technology.

Elements	Appropriate technology
Basic entities acknowledged	Liberal market
	Small and low capital technology
	Environment
	Others: NGOs, the national government, international cooperation agencies, intergovernmental agencies and financial institutions
Assumptions about natural in relationships	Hierarchy
	Inclusiveness
	Nature, especially wind, sun, water and biomass, is seen as the core for appropriate technology generation
	Small scale technology less harmful for the environment
Agents and their motives	National NGOs: support small scale producers, and small agro-industries to implement demonstrative projects
	International cooperation agencies: Moral mandate and market opportunity
	Small and micro scale enterprises: increase income generation, productivity and access to technologies
	National government: appropriate technology for poor people is not a priority
Key metaphors and other rhetorical devices	Rhetorical devices: 'small is beautiful', 'self-reliance', 'poor'

subsistence and not fulfilling anymore the current needs (Sanchez, 2009, interview). For instance, discourse subscribers would suggest the use of a solar cooking or improve the efficiency of stoves for cooking instead of using wood. According to their proponents, appropriate technology allows SMEs to increase income, efficiency and productivity by providing higher added value to their products. Access of production technologies, information and communication technologies, technologies for irrigation and organic production, and building small power plants are seen as key technologies to help people in poverty (ITACAB website, 2011; ITDG, 2009).

The environment is an acknowledged entity. Discourse proponents argue that the scale of production and technology matters for the level of impact on the environment. The discourse points out the sustainable features of small scale technologies. Oppositely, large scale technology and mass production are seen as more harmful for the environment and less suitable for most parts of the world. Therefore, the use of appropriate technology by small scale technologies is seen the best way to protect the environment.

Other entities acknowledged are NGOs, international cooperation agencies, intergovernmental agencies, international financial institutions and national government.

Assumptions about natural relationships

A hierarchy is perceived, based on having or not having modern technological devices. The discourse allocates enterprises in two clusters based on their capacities to develop and handle technology. The categorization is based on a bipolar worldview with hierarchical relationships. In one corner, the wealthy, powerful, modern and large scale enterprises are allocated, and in the other corner the poor, powerless, traditional practices and small scale enterprises. This is clearly established in the following quote of Ernst Fritz Schumacher: 'a new approach is needed to aid and development. This approach will be different to the older in the sense that it takes poverty seriously. It is wrong to say: What is good for rich people is good for poor people as well' (Soluciones Practicas website, 2011).

According to the appropriate technology discourse the only way to overcome this polarity is by transferring and adapting modern technologies from developed countries to developing countries according to local conditions. Also, the improvement of certain local traditional technologies might be helpful to overcome polarity (ITDG, 2009). The discourse assumes that elites (e.g. international cooperation agencies) with agency capacities are in a position to generate and transfer technologies. Such advantaged elites are in power or moral mandate to provide help to the marginalized people. Proponents might claim that the appropriate technology discourse is a genuine grassroots solution, but opponents argue that connecting local communities with the modern world by technology devices erodes traditional life-style.

Inclusion of poor people in modernity is based on the ownership of technology. In the discourse, inclusion means to make available technological devices for excluded poor people so that they can generate enterprises, improve production processes and benefit more from liberal markets (ITDG, 2009). The inclusion is based on self-reliance of people rather than dependency on standardized technologies (Sanchez, 2009, interview). The ownership of technology by small and micro scale enterprises is central for business growth and sustainability.

Regarding the relationship with nature, the appropriate technology discourse claims the efficient use of natural resources and diminishing environmental impacts. Particular attention is given in the discourse to technologies that use renewable energies such as wind, sun and water as a strategy to mitigate climate change (ITDG, 2009).

Agents and their motives

Agency is granted to NGOs with technological expertise, including traditional and technocratic NGOs. Most NGOs with agency capacities are NGOs operating at the national level. NGOs subscribing to the appropriate technology discourse claim to work in favor of 'underprivileged' and 'poor' people, and fight injustice. NGOs aim to set and upgrade small and micro scale enterprises oriented to local and national markets by implementing demonstration projects (IPES website, 2012; ITACAB website, 2011). NGOs endorsing the discourse of appropriate technology provide capacity building on low cost technologies using participatory approaches, and influence policy makers to create public polices to support appropriate technology application (ITDG, 2009). The ultimate aim of the appropriate technology agents is to fight poverty in developing countries (ITACAB website, 2011; ITDG, 2009; Sanchez, 2009, interview).

Agency is also granted to international cooperation agencies, including international NGOs, intergovernmental agencies and governmental agencies of developed countries. The research centers and consultancy firms on technology that work directly with technocratic NGOs are also included. International cooperation agencies provide funding, policy guidance and technological expertise to NGOs and national governments to implement demonstration projects on appropriate technology. They also facilitate the transfer of technology among countries and regions worldwide in cooperation with national agents. Moral mandates, environmental concerns and the creation of market opportunity for small scale technology devices are the main motivations of global agents (ITDG, 2009).

Agents include micro and small scale enterprises, regardless their type and economic sector (Sanchez, 2008). Agents might be small scale producers (e.g. micro and small crop producers), small scale industries (e.g. food-industries, stockbreeding and breeding of Andean camelids), small scale workshops and entrepreneurs (ITACAB website, 2011; Su, 2004; ITDG, 2009). The main motivations of SMEs to subscribe to this discourse are increase of income generation, productivity and access to technology in order to be(come) part of modernity.

The national government has agency but does not prioritize appropriate technologies. The national government is seen as biased to promote mainly up-to-date modern technologies. In the eyes of the national government, appropriate technology will bring us backward. Moreover, the promoters of appropriate technologies are seen as actors against economic development of the country. Therefore, governmental agencies and NGOs have scarce cooperation. The national government prioritizes larger scale industries and exports of both raw materials and commodities to international markets, neglecting the value of appropriate technology for the national economy, especially for poor people (Sanchez, 2009, interview).

Key metaphors and their rhetorical devices

The discourse of appropriate technology includes several key rhetorical devices such as the terms 'small is better', 'help themselves', 'self-reliance' and 'poor'. However, the most emblematic rhetorical device that embodies appropriate technology is 'small is beautiful'. According to the discourse proponents, the term 'small is beautiful' not necessarily means small in size nor in scale, but small in the decentralized features of technological solutions across the world.

7.3.4 Conclusions

a. Based on empirical research in Peru, the discourse analysis permits us to identify two discourses of sustainable production: the cleaner production discourse and the appropriate technology discourse. The two discourses have communalities and differences. The communalities are that both discourses endorse the central role of technology to foster sustainability of SMEs. Sustainability of SMEs is understood as the increase of income generation, efficient use of natural resources and further productivity. The discourses rest on hierarchical worldviews of society and technocratic views of nature. Both discourses identify NGOs, (inter)governmental agencies, international cooperation agencies and SMEs as key actors. The difference between both discourses is that while the cleaner production discourse highlights the access to capital intensive modern technology, the connection of SMEs with global markets and the re-enforcing character of environmental improvement and higher economic benefits, the appropriate technology highlights the development of low capital rural small scale technology, the adjustment of technology to local conditions and the self-reliance of SMEs on local markets.

b.   The cleaner production discourse and the appropriate technology discourse acknowledge the central role of the liberal market for improving the performance of SMEs. The liberal market is seen as a powerful force for business competitiveness. While the liberal market is seen in the cleaner production discourse as a necessary condition for improving environmental conditions of SMEs, the appropriate technology discourse sees the liberal market either as opportunity or threat. The liberal market is an opportunity as long as SMEs own the technology. Otherwise SMEs will be incapable to succeed.

c.   The improvement of environmental conditions in SMEs according to both discourses is mainly oriented to business growth and access to profitable markets. However, both discourses have differences in the way to reach such improvement. The cleaner production discourse argues that the best way to reduce the environmental impact of SMEs is increasing efficiency in their production processes, regardless of the type, size or economic sector. The appropriate technology discourse prefers the use of small scale cleaner technologies and adjusts them to local conditions to reduce the environmental burden of SMEs.

d.   The cleaner production discourse and the appropriate technology discourse are rich in rhetorical devices. While eco-efficiency is the device located at the core of the cleaner production discourse, 'small is beautiful' is the most emblematic rhetorical device that embodies appropriate technology. While the rhetorical devices of the cleaner production discourse describe the engagement of SMEs to profitable markets and the centralized way to approach sustainable production, the rhetorical devices of the appropriate technology production discourse describe the self-reliance of SMEs and the decentralized way to approach sustainable production.

7.4 Discourses of sustainable production in a two-dimensional policy realm

A two dimensional figure has been constructed to allocate the two discourses of sustainable production (Figure 7.2). The figure consists of the social (axis A) and economic (axis B) realm. On the vertical axis (A) the focus is on the social foundation for structuring society, whereas the horizontal axis (B) reflects the economic foundation for organizing society.

The extreme poles of axis A represent 'individualistic' and 'communitarian' values. 'Individualistic' values highlight the individual interests over collective ones and a faith in the capacity of individual action and ambition to create wealth and to bring about progress. The 'communitarian' values highlight the collective values and common identity of people within a shared geographical scope from local to global community. In the first case, individual interests are the base for relationships among actors. This pole of the axis refers to individual or group benefits rather than collective welfare of the whole society. In the second case, the social strengthening of local communities is central. The collective values are among others solidarity towards people in need, sense of community and common cultural identity.

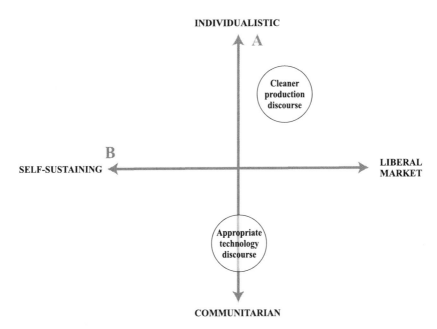

Figure 7.2. Discourses of sustainable production in two dimensional axes.

The extreme poles of axis B represent 'self-sustaining' and 'liberal market' values. 'Self-sustaining' values refer to the state of not requiring any outside aid, support, or interaction, for survival and autonomy. 'Liberal market' values refer to the mainstream free market economy where business has a central role in production and commercialization of goods and services at the global level. In the first case, discourses are oriented to meet the needs of technology and markets of SMEs in a decentralized way. Technology is developed in close cooperation with individual SMEs or clusters of SMEs. In the second case, discourses are oriented to meet the needs of technology and markets of SMEs in a centralized way. Liberal market values put the globalization and connectedness of actors from local to global scale central (Figure 7.2).

Now, the discourses identified are allocated in the two dimensional matrix (Figure 7.2). The cleaner production discourse is allocated in the crossing of 'individualistic' values and 'liberal market' values. The appropriate technology discourse is rooted strongly on 'communitarian' values and it links 'self-sustaining' and 'liberal market' values.

7.5 Major patterns and trends: networks and discourses

In this section networks and discourses are put together in a single framework (Figure 7.3) to understand the main patterns and trends resulting from their relationships.

The main characteristics identified in the relationships between networks and discourses of sustainable production are the following: (1) sustainable production networks are grouped according to the discourses subscribed; (2) the embedding of the liberal market in the national economy is triggering the increase in scale of micro and small scale enterprises; (3) SMEs

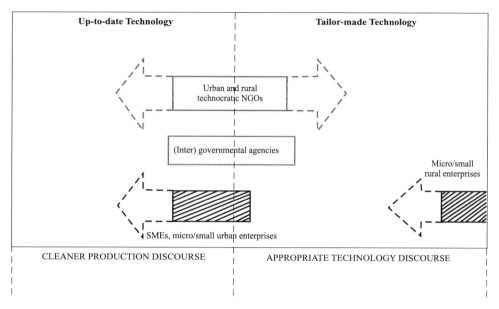

Figure 7.3. Main discourse and network patterns and trends of NGOs, (inter)governmental agencies, SMEs and micro scale enterprises.

subscribing to the sustainable production discourses are diverse in size and market target; and (4) new roles of NGOs are emerging.

The sustainable production networks identified subscribe either to the cleaner production discourse or to the appropriate technology discourse. The eco-efficiency network, the cleaner technology network, the urban cleaner production and the technological innovation network endorse the cleaner production discourse. The appropriate technology network endorses the appropriate technology discourse. The main difference between both discourses has to do with the type of technology preferred to make SMEs sustainable, but is in the end about world views. While modern up-to-date technology is the right option for the cleaner production discourse's proponents, intermediate technology is the correct option for the appropriate technology discourse's proponents.

Second, actual rural micro and small scale enterprises are gaining scale as an answer to the market demand for their products. The increase in scale is not only in individual businesses but also in cooperatives, clusters of rural enterprises and rural enterprise associations. This development is happening not only following demands in metropolitan cities like Lima but also in intermediary cities like Trujillo, Arequipa, Puno and Huancayo. It is important to notice that the increase in scale implies the building of organization structures such as SME platforms. This development might be good in order to reduce the current dispersion of SMEs. However, the increase in orientation of rural small scale enterprises towards profitable markets might impact the capacity of rural communities to be self-reliant in local food supply.

Third, SMEs subscribing to the cleaner production discourse and the appropriate technology discourse are diverse in type and market targets. While cleaner production is more feasible to apply in SMEs oriented to competitive markets, appropriate technology is more feasible to apply in small and micro scale enterprises oriented to local markets. Subscribers of the cleaner production discourse are eager to become exporters and to expand to new markets. In the views of the discourse proponents, the integration of SMEs in global markets enhances the adoption of eco-efficiency, environmental friendly technologies and innovation in production processes. Subscribers to the appropriate technology discourse include micro and small enterprises that supply small markets, particularly in intermediary cities, rather than SMEs supplying larger markets. Most small scale enterprises subscribing to the appropriate technology discourse are based on locally available natural resources and expertise. Usual buyers of SMEs subscribing the appropriate technology discourse are individual local consumers, local governments and larger extractive multinational companies.

Fourth, new types of NGOs found in the sustainable production networks are technocratic NGOs and government organized non-governmental organizations (GONGOs). Although all NGOs involved in the sustainable production networks have expertise on technology, technocratic NGOs are highly specialized in engineering processes and GONGOs in policies on technology. It is important to notice that technocratic NGOs with longer histories such as IPES have evolved from conventional NGOs. However, other technocratic NGOs such as CER, and GONGOs such as ITACAB and CONCYTEC, have emerged later as a result of the expansion of the sustainable production networks. While conventional NGOs usually make broad capacity building on device handling, technocratic NGOs are currently becoming highly specialized thinktanks on sustainable production. Therefore, new roles of NGOs include the promotion, transfer, adjustment and development of sustainable technologies oriented to make production of SMEs more sustainable.

The main trends identified from the relationships between networks and discourses of sustainable production are two. First, NGOs are getting closer to SMEs by providing them with specialized services for the improvement of the economic and environmental performance of their production processes. NGOs aim to cover rural enterprises and urban enterprises, micro scale enterprises and medium-sized enterprises, and local market oriented SMEs and international market oriented SMEs. Due to the diversity of SMEs in their size and market targets, NGOs are starting to offer tailor made services for SMEs. For instance, IPES focuses on urban small scale enterprises, CER and CITEs focus on medium scale enterprises and Practical Action Peru focusses on rural small scale enterprises. Second, the better positioning of SMEs as suppliers of larger domestic companies and of international buyers seems to increase their interest in cleaner production. This new trend might be because under the liberal market frame the cleaner production discourse looks more appealing for SMEs than the appropriate technology discourse. Being a supplier of value chains seems to motivate SMEs to move from 'indifference' to 'concern' for cleaner production measures. Additionally, local SMEs, especially rural small and micro enterprises, will be forced to consider the appropriate technology perspective more seriously as the environmental crisis worldwide gets deeper.

7.6 Conclusions

a. The sustainable production networks involve not only NGOs, but also (inter)governmental agencies as key actors. NGOs identified in the networks are technocratic NGOs working in rural and urban areas. Although CONCYTEC and ITACAB are (inter)governmental agencies, they perform similar roles in the networks as NGOs. (Inter)Governmental agencies are not NGOs, but operate as government organized non-governmental organizations (GONGOs). The following roles for NGOs and GONGOs are identified: promotion, transfer, adjustment and development of sustainable technologies oriented to make production of SMEs more sustainable. Together with the sustainable production main promoters such as UNIDO, UNEP, State Secretariat for Economic Affairs of the Swiss Federation, the Inter-American Development Bank, the Organization of American States and Practical Action International, the national key actors such as CER, CONCYTEC, ITACAB, Practical Action Peru, IPES and CITEs, intend to change production processes of SMEs towards more environmental friendly and socially inclusive production. NGOs and (inter)governmental agencies endorse both the cleaner production discourse and the appropriate technology discourse. The case study shows that key actors in the networks are dependent not only on funding and expertise from international cooperation agencies but also on 'will'. The sustainable production networks lack committed 'self-mobilizing' key actors. This makes NGOs and (inter)governmental agencies failing in achieving their aims of expanding sustainable production.

b. Networks of sustainable production show scarce cooperation of actors, notwithstanding the strong commonalities in their views. Sustainable production networks are characterized by scarce cross cooperation, high duplication of efforts and low effectiveness. SMEs have low trust in the agents of sustainable production discourses, especially NGOs. Although technocratic approaches that dominate in the sustainable production networks are key to ensure the environmental adjustment and to change SMEs production processes, they are not enough to move SMEs towards sustainability. Technocratic views in the sustainable production networks have to evolve towards socio-technical views. Greater specialization in sustainable technologies is not enough to engage SMEs in sustainable production networks and discourses. The majority of SMEs in Peru, usually carrying environmentally inefficient production processes, is surviving in tiny urban markets, and are dispersed and fragmented in hundreds of small tiny associations and single enterprises. Therefore, awareness of the social and cultural background of SMEs and understanding the SMEs' motivations are necessary to persuade SMEs and to increase their participation in the networks.

c. Power of SMEs, especially of small- and micro-scale enterprises, is not fully acknowledged by agents in the sustainable production discourses. SMEs show power by accepting or denying connect to the sustainable production networks, and by subscribing or being indifferent to the discourses. This power rests on the capacity of SMEs to be anchored within the local social networks of production and service provision. The scarce motivation of SMEs to buy-in cleaner production or appropriate technology put them in a powerful position to delegitimize interventions of sustainable production promoters. This 'unenthusiastic' attitude of SMEs does not mean that they do not need support. SMEs need support, especially frontrunner SMEs, to overcome constrains in terms of business capacities and access to profitable markets. The case study also shows that agents of the appropriate technology discourse are more aware of the power of SMEs than agents of the cleaner production discourse. Small and micro scale enterprises interested in adjusting and innovating technology to their needs have higher chances to participate in the decision making of agents. However, the sustainable production networks lack platforms or other institutionalized hubs for coordination between agents and

SMEs. The acknowledgement of the power of SMEs passes by realizing that SMEs are not just beneficiaries in the networks but legitimate actors. Without the participation of SMEs in the design of intervention policies, the good-will of sustainable production promoters will not move beyond a few 'successful' experiences of sustainable SMEs. NGOs need to acknowledge the motivations of SMEs, be embedded in SMEs' local social networks and establish the conditions for SMEs to show their social power at the market.

d. Market forces are not enough to push SMEs to the adoption of cleaner production and appropriate technologies. Certainly global markets play a valuable role in fostering the change of SMEs from pollution to sustainability, especially in increasing the interest of SMEs for cleaner production. However, sustainable production networks are failing in their aims to expand the number of SMEs with sustainable practices. This is so because the networks are applying a passive approach of waiting for the voluntary 'good-will' of SMEs or the 'invisible hand' of the market to bring SMEs towards sustainable production. A more proactive approach to expand the number of SMEs with sustainable practices, regardless of type, size or market, is needed. Improving the environmental performance of production cannot be a privilege of some SMEs but needs to become a compulsory condition to operate legitimately as a business for all SMEs. So, the widening of sustainable production in SMEs needs the strengthening of SME associations and of clusters, and an active promotion role of the national and local governments.

Chapter 8.
Comparative analysis and theoretical reflections

8.1 Introduction

In this chapter the discourses and networks promoting sustainability of SMEs in Peru are analyzed comparatively. Based on this analysis, the roles and perspectives of NGOs are investigated in relation to the Peruvian government and to international cooperation agencies. Furthermore, trends for the future are explored. Finally, a theoretical reflection on the outcomes of the research is presented against the backdrop of current debates on network society theory and ecological modernization theory. The reflection focuses on issues such as network configuration, cooperation, power relationship, institutional transformation, environmental rationality and market.

8.2 Discourses

The case studies of organic production, business social responsibility and sustainable production in Chapter 5, 6 and 7 address seven discourses of sustainability of SMEs. As it has been analyzed in these chapters, business social responsibility and sustainable production have evolved in two discourses, and organic production has evolved in three discourses (Table 8.1).

Although all discourses in the case domains listed in the Table 8.1 focus on sustainability of SMEs, there are differences in their perspectives. Differences in the discourses depend on the fact that certain values prevail over others. Based on the case studies we identify four value clusters that are central to the discourses. These clusters can be labeled as 'individualistic' values, 'communitarian' values, 'self-sustaining' values/'rights' values and 'liberal market' values. In the market access discourse individual gains are central rather than collective welfare, in the market adaptation discourse solidarity is a communitarian value that mediates the relationship among actors, in the appropriate technology discourse self-reliance values such as independency, autonomy and trust in local resources are put first, in the corporate responsibility discourse 'liberal market' values such as competitiveness, mobilization of goods and services and interlink of actors at global level are central and, finally, the business upgrading discourse is based on values that protect the social and economic rights of vulnerable people from powerful self-interested people. Each domain also carries several divergent discourses. For instance, organic production is a ground for three competing perspectives about what organic production means and how SMEs should

Table 8.1. Discourses that promote sustainability of SMEs.

Organic production	Business social responsibility	Sustainable production
Market adaptation discourse	Business upgrading discourse	Cleaner production discourse
Market access discourse	Corporate responsibility discourse	Appropriate technology discourse
Market democratization discourse		

be supported; the market adaptation discourse, the market democratization discourse and the market access discourse. The access discourse diverges strongly from the rest of the discourses. The adaptation discourse and democratization discourse are much closer, despite the differences. While the discourse of market democratization emphasizes the self-sufficiency values and the discourse of market adaptation emphasizes 'communitarian' values, the discourse of market access emphasizes 'individualistic' values. Business social responsibility encompasses two divergent discourses about what type of SMEs should be prioritized. While the corporate responsibility discourse strongly shares perspectives with the liberal market discourse and it is catching up with social concerns, the business upgrading discourse shares perspectives with the rights discourse and it is catching up with market rationality. Business social responsibility discourses are limited to SMEs engaged with global markets. However, in both discourses the change of SMEs from 'business as usual' to more sustainable production is emphasized. Sustainable production also encompasses two deviating discourses in their perspectives about how to make SMEs sustainable and what type of SMEs should be prioritized. The two discourses are clearly two different perspectives to cope with SMEs. Each discourse acknowledges a particular segment of SMEs; the cleaner production discourse focuses on medium-sized and urban enterprises and the appropriate technological discourse focuses on small and rural enterprises. While the discourse of cleaner production strongly shares perspectives with the discourse of liberal market, the discourse of appropriate technology strongly shares perspectives with the discourse of self-reliance.

Based on the aforementioned values, the discourses of sustainability of SMEs can be grouped in four groups. The first group, based on 'individualistic' and 'liberal market' values, includes the market access discourse, the cleaner production discourse and the corporate responsibility discourse. The second group, based on 'communitarian' values, includes the market adaptation discourse. The third group, based on 'self-sustaining' values, includes the market democratization discourse and the appropriate technology discourse respectively. Finally, the 'rights' values are central in the business upgrading discourse.

The discourses of sustainability of SMEs not only hold differences as explained above but also similarities. Similarities in discourse storylines are drawn from the relationship between discourses and values. As shown in the Figure 8.1 the discourses of sustainability of SMEs can be mapped over two value axes. Figure 8.1 puts together the Figure 5.2, the Figure 6.2 and the Figure 7.2. The Y axis represents the difference between 'individualistic' and 'communitarian' values. These values can be regarded as mutually opposing. The X axis represents the difference between a 'self-sustaining'/'rights' orientation and a liberal market orientation. On this axis, the self-sustaining and the right value clusters are both at one extreme, in opposition to an orientation that puts a high value on the liberal market. The place of the discourses on the axis is based on the analysis in the previous chapters. When a discourse is positioned in the middle, this means that it encompasses values from both ends of the axis. Similarities identified are the following: first, the allocation of the adaptation discourse in the value axes means that the discourse storyline is widespread between the central role of business for economic structuring of society and the social strengthening of local communities based on collective values, sense of community and cultural identity. Second, the allocation of the access discourse, the corporate responsibility discourse and the cleaner production discourse in the value axes means that their discourse storylines are extended between the interest and action of individuals to create wealth and to bring about

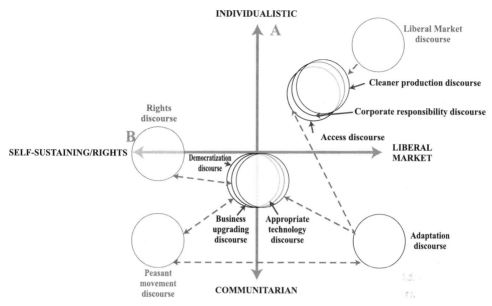

Figure 8.1. Discourses mapped over two values axes.

progress, and the central role of business in production and commercialization of goods and services at a global level. Finally, the allocation of the democratization discourse, the business upgrading discourse and the appropriate technology discourse in the value axes means that their discourse storylines are stretched between the autonomous supply for self-consumption of the family, the community or the country and the social strengthening of local communities based on collective values, sense of community and cultural identity.

Another important similarity is found in the origins of the discourses. As I have described in the previous chapters, the discourses of sustainability of SMEs have evolved from long-standing antagonist discourses: the liberal market discourse on one hand and the social movement discourses on the other hand. Social movement discourses are against or resist liberal market. Their proponents usually blame the state and companies, and hold adversarial perspectives and reactive positions against larger and transnational companies. The liberal market discourse focuses on economic rationality, marginalizing or neglecting environmental rationality. Their proponents consider central economic growth and global free trade.

As shown in the Figure 8.1 the liberal market discourse and the social movement discourses are considered 'mother' discourses of the discourses of sustainability of SMEs. In the social movement discourses three different versions can be distinguished: the peasant movement discourse, the rights discourse and the self-sustaining discourse. The peasant movement discourse is the 'mother' discourse of the market democratization discourse, the market adaptation discourse and the market access discourse. Being more precisely, the democratization discourse has evolved from the adaptation discourse but it has got influence from the peasant movement discourse as well. The market access discourse has also emerged from the adaptation discourse. The adaptation discourse has its origin in the peasant movement discourse. The rights discourse is the 'mother' discourse of the business upgrading discourse and the 'self-sustaining' discourse of the appropriate technology discourse. The liberal market discourse is the 'mother' discourse of the cleaner production discourse and corporate responsibility discourse.

This relationship in their origins between the 'offspring' discourses with their 'mother' discourses makes that the discourses of sustainability of SMEs share views with the social movement discourses or the liberal market discourse. While the discourses of market access, cleaner production and corporate responsibility strongly share perspectives with the discourse of liberal market, the discourses of market adaptation, market democratization, business upgrading and appropriate technology strongly share perspectives with the discourses of social movement.

Another common pattern of the seven discourses identified is that all accept the reality of the liberal market. The liberal market is seen as opportunity or a battle field that cannot be avoided. Discourses of sustainability of SMEs do not encompass resistance or adversarial perspectives on the liberal market. For instance, storylines of the organic production discourses extend from accepting the free market as an unavoidable matter of fact to positively endorsing it (shown in Section 5.3.2).

As Figure 8.2 illustrates, the main line of tension in the discourses is that between the peasant movement and rights discourses on one hand, and the liberal market discourse on the other hand. Depending on their position on this line the seven discourses can be clustered in two main groups, which can be labeled as the 'market justice' discourses and the 'sustainable market' discourses (Table 8.2). The names underline the main features of the groups. In the market justice discourses communitarian values are more central, and emphasis is put on inclusion of emerging new economic actors under the liberal market frame.

The market justice discourses and the sustainable market discourses are not only different in their position towards social movement and market, but also in their interpretation of environmental reform and sustainability. In the market justice discourses, environment is seen as something to be protected, defended and stood up for. In the sustainable market discourses emphasis is put on successful market performance. Environment is seen as business opportunity and a legitimacy factor by their proponents.

Justice and sustainability, the concepts emphasized in the discourses of sustainability of SMEs, are also underlined in the notion of sustainable development. For instance, authors such as Langhelle (2000) states that social justice – understood as need satisfaction – is at the core of

Table 8.2. Clusters of discourses of sustainability of SMEs.

Discourses of sustainability of SMEs

Market justice discourses	Market adaptation discourse
	Market democratization discourse
	Business upgrading discourse
	Appropriate technology discourse
Sustainable market discourses	Market access discourse
	Cleaner production discourse
	Corporate responsibility discourse

sustainable development, and sustainability is considered as aim in the long-term development. Although justice and sustainability concepts are central in both the discourses of sustainability of SMEs and the notion of sustainable development, they are not similarly approached. While in the sustainable development notion, justice and sustainability are seen as equal interdependent elements (Langhelle, 2000), the discourses of sustainability of SMEs establish more weight to one of them. This difference in weight has to do with the difference in their values based on and the distinct ways of understanding the environment. Moreover, while sustainable development refers to social justice and environmental sustainability in broader terms, the discourses of sustainability of SMEs refer to justice and sustainability within the boundaries of the liberal market. Thus, the separation of the discourses of sustainability of SMEs in two well defined groups shows the tension between justice and sustainability rather than a harmonious set of ideas as they are usually presented in the notion of sustainable development.

8.3 Network and discourse relationships

The study shows that new discourses are not only born and nested in networks, but also contribute to the development of new networks. Moreover, networks and discourses are not necessarily linked in a one-to-one relationship over time. Networks can endorse more than one discourse and one discourse can be shared by several networks. For instance, NGOs engaged in the sustainable production networks subscribe either the cleaner production discourse or the appropriate technology discourse. The macro discourse of 'sustainable market' is shared by NGOs engaged in the networks of organic production, business social responsibility and sustainable production.

The general relationships between networks and discourses, as they are described in the previous chapters, are visualized in Figure 8.2. In the Figure, discourses are considered a static body of thoughts ordered in blocks where networks unfold along the discourses. The discourses are clustered according to the Table 8.2. The arrows and blocks represent five processes of change in the relationships between networks and discourses of sustainability of SMEs taking place currently.

First, changes are to be found in the networks from adversarial to collaborative views. The relationships between networks and discourses can be understood in terms of ties and cooperation. This pattern of changes is unfolded by the organic production networks and is displayed as a 'unidirectional shift' from the discourses of social movements towards the discourse of liberal market. Although the organic production networks diverge in perspectives, they have strong ties. For instance, the agro-ecological network and the ecological farming network endorse 'market justice' and the organic market network endorses 'sustainable market'. The supremacy of cooperation does not mean absence of conflicts. The three networks of organic production are engaged in conflicts, but nonetheless, they cooperate intensively in supporting SMEs.

Second, we can observe changes in the networks from solidarity to business performance due to the more central role of the liberal market. This change runs parallel to the aforementioned change. In the 1980s and 1990s networks were driven by solidarity. This has been the main feature of the networks of long history. For instance, the agro-ecological network, the ecological farming network and the social justice network, even of the organic market network during its first years. From 2000 onwards solidarity is being replaced by the measuring of performance. This new development, which can be clearly observed in the organic market network, the business network

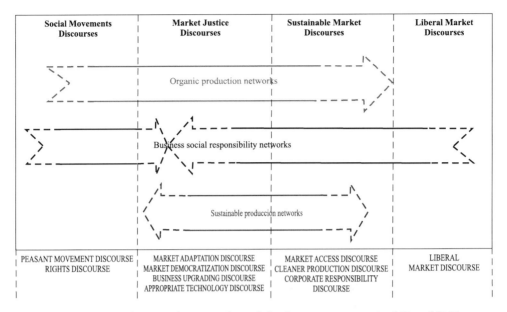

Figure 8.2. Relationship between the networks and the discourses on sustainability of SMEs.

and the eco-efficiency network, has to do with the more central role of the liberal market discourse in the networks. Extensive evidence has been provided in Chapter 5, Chapter 6 and Chapter 7.

Third, changes can be witnessed in the networks towards a convergence of views. In the case of business social responsibility, the social justice network and the business network do not share collaborative ties but they converge in their views about SMEs. Both networks agree on the inclusion of emerging small business in the market, for instance, by upgrading production to meet buyer requirements and engaging SMEs in supply chains. What is significant regarding these converging views is that both networks come from radical opposing discourses: the rights discourse and the liberal market discourse. The fact that the networks currently endorse two different discourses – the upgrading discourse or the corporate responsibility discourse – has to do with emphasizing either rights or liberal market views and prioritizing in their interventions either SMEs or larger companies, but both networks consider central the role of sustainable value chain to improve the situation of small scale enterprises. Moreover, the 'mother' discourses may still play a role in the business social responsibility networks by 'shadowing' the converging views of actors. These peculiarities of the business social responsibility networks contribute to their isolation from each other rather than explore the aforementioned converging perspectives about SMEs and look for cooperation. Thus, the business social responsibility domain is characterized by parallel networks with converging discourses. This is depicted in Figure 8.2 as a 'converging shift' of networks.

Fourth, changes can be noticed in the networks towards a divergence of views. In the case of the sustainable production networks, they are tied to each other but there is divergence in their views about how to cope with SMEs, especially between the four networks subscribing to the sustainable market discourses and the one subscribing the market justice discourses. This is depicted in Figure 8.2 as a 'diverging shift' of networks towards the discourses of social movement and the discourse of liberal market. Sustainable production networks are more diverse and heterogeneous than both the organic production networks and the business social responsibility networks. The networks are able to cope with market and social concerns of SMEs but the differences in their discourses do not favor further cooperation to support SMEs. In other words, the presence of ties among actors in the networks is not a guarantee for cooperation. Cooperation is scarce and duplications of efforts are strong among the networks of sustainable production. This failing characteristic of the networks of sustainable production, particularly of the cleaner technology network, has been pointed out by UNEP (2010) in an evaluation of the technological transfer platform (TTN) at global level.

Fifth, new types of NGOs emerge to fulfill market demands. The emergence of the discourses on sustainability of SMEs has motivated the establishment of new types of NGOs such as market NGOs, producer NGOs, business NGOs and technocratic NGOs (Table 8.3). More diversity of NGOs is found in organic production and sustainable production. As a result of this diversification of NGOs, for instance in the case of organic production, the struggle for leading positions in the platforms and the competition for scarce funding have also intensified. While conventional NGOs are the NGOs with a longer history, producer and market NGOs are newcomers. Both market NGOs and business NGOs provide services for economic actors. While market NGOs focus on small business, business NGOs focus on larger companies. This diversification of NGOs is to fulfill the business growth and market demands of small producers. So, the market pushes NGOs to change their priorities and approaches regarding SMEs. For instance, while conventional NGOs are under pressure to adopt the market approach, market NGOs are created as an answer to the market needs of small organic producers. In addition to the new types of NGOs, (inter)governmental agencies have been shown to perform as key actors in the sustainable production networks. (Inter)Governmental agencies operate as government organized non-governmental organizations (GONGOs). Like NGOs, government organized NGOs (Table 8.3), promote sustainability of SMEs, compete for funding and operate projects of international cooperation agencies. This means that in the sustainable production networks (inter)governmental agencies work closely with NGOs, especially with technocratic NGOs who share their views. The performance of government organized NGOs in the networks has been found out to be less effective in promoting sustainability of SMEs than more typical NGOs (see Chapter 7).

8.4 Future trends

The shifts identified in this study, schematically summarized in Figure 8.2, not only help to understand the current dynamics of the networks and discourses of sustainability of SMEs but may also shed some light on tendencies for the future.

A major tendency found in the current dynamics, and likely to extend into the future, is the embedding of liberal market rationality into all networks. This embedding is also triggering the diversification of NGOs and the emergence of new roles. For example, we see NGOs run by producers, NGOs oriented to business and NGOs performing intermediary roles on the market, among other types of NGOs and roles. Comparable findings are reported by Beloe, Elkington, Fry Hester, Newell, Kell and Aloisi (2003) highlighting a more central position of NGOs in the market and an increasing importance of the market for the sustainability of NGOs. At the same

Table 8.3. Types of NGOs identified and the discourses subscribed.

Domain	Discourse	Types of NGOs
Organic production	Market adaptation	conventional NGOs
	Market access	market NGOs
	Market democratization	producer NGOs
Business social responsibility	Business upgrading	conventional NGOs
	Corporate responsibility	business NGOs
Sustainable production	Cleaner production	technocratic NGOs
		conventional NGOs
		government organized NGOs
	Appropriate technology	technocratic NGOs
		government organized NGOs

time, however, it is plausible that networks endorsing social movement discourses will not fully abandon them, as is illustrated by the organic production networks.

Another tendency that may well extend into the future is that actors that share the same views sooner or later are likely to come together. We might expect this, for instance, in the business social responsibility networks where there are two well defined networks of actors working separately despite their shared views regarding SMEs. The motivation of SMEs for reaching sustainable markets represents a driving force for cooperation. In this process of cooperation between larger companies and SMEs, NGOs have played the role of intermediaries. Furthermore, the common perspectives of the discourses of sustainable production, organic production and business social responsibility regarding to environment and market are potential bases for cooperation. For instance, the organic production networks and the business social responsibility networks need knowledge about cleaner and appropriate technologies to achieve their aims towards SMEs and the sustainable production networks need to widen their scope of SMEs to gain legitimacy. This implies that sustainable production networks may become 'bridges' between organic production and business social responsible networks.

The tendency of the social and business spheres in the networks is to get blended due to the social and economic roles of SMEs. SMEs have traditionally hold strong connections with local social networks. But now, SMEs are moving from local social networks to global economic networks. This further connection with global economic networks is causing, on the one hand, NGOs to move from social to economic roles and, on the other hand, larger companies to move from economic to social roles. This blending is currently happening in the organic production networks to a major degree and to a minor degree in the rest of the networks.

An additional tendency is the increase of service NGOs and consultancy like organizations. A further increase of service NGOs (e.g. market NGOs, business NGOs, technocratic NGOs and producer NGOs) aimed to support SMEs for market access can also be expected for the future. Collaborating with market actors, service NGOs are becoming valuable actors to support the market access of SMEs. Service NGOs are market driven and prioritize SMEs that hold capital, technology and managerial capacities. Similar findings are reported by Diaz-Albertini (2001) in the sense that NGOs are becoming professional organizations, leaving activism against the state. While several NGOs are quite critical of the government, no NGOs are anti-governmental anymore. Some of them even are pro-governmental (e.g. government organized NGOs) and pro-business (e.g. business NGOs) as they work closely with governmental agencies and business in their developments tasks (see also the next section). Furthermore, NGOs are not anymore an organization emphasizing activism and adversarial positions. NGOs and business are moving to establish ties and eventually cooperation as globalization bring them closer. This means that NGOs are undergoing a process of change of identity from politics to 'technique' and 'efficiency'. According to Diaz-Albertini (2001) this generates a tension in NGOs: on the one hand NGOs become project operators and on the other hand they may easier fall into political clientelism.

The last main tendency identified is that small businesses are gaining bargaining and decision power in the networks. The case studies provide evidence of the changes in the position of SMEs in the networks in terms of gaining bargaining and decision power in the local and national platforms. This is particularly happening in the organic production networks. SMEs are gaining recognition as actor in the networks. For instance, in the networks of organic production rural SMEs are shifting from beneficiaries to active participants. Rural SMEs are also getting emancipated from conventional NGOs. However, most SMEs connected to the networks of sustainable production and the networks of business social responsibility are more passive in their participation and indifferent to NGOs.

8.5 Roles of government and international cooperation agencies

8.5.1 The Peruvian government

Several voices have been identified with regard to the role of the Peruvian government in the sustainability of SMEs. While proponents of the market access discourse and the corporate responsibility discourse label the government as incapable and ineffective, proponents of the market democratization discourse, the market adaptation discourse, the business upgrading discourse, the cleaner technology discourse and the appropriate technology discourse label the government as careless, unfair and irresponsible. These findings provide evidence that local NGOs have low trust in the national government. This mutual mistrust between NGOs and the government in Peru has been pointed out in other literature as well, for instance by Diaz-Albertini (2001) in a study about the political participation of NGOs in Peru. According to Seegers (2008, interview), the government as institution has little or no cooperation with the networks that support SMEs in Peru. When there is cooperation, NGOs are usually limited to operate government funded projects, for instance, the INCAGRO program oriented to agricultural projects. If so, most of the cooperation of governmental agencies with the networks of sustainability of SMEs is in terms of personal motivation rather than institutional commitment, as shown the Section 5.2.1. Only in the cleaner technology network, the appropriate technology network and the technological innovation network (inter)governmental agencies perform central roles and work closely with NGOs. However, cooperation in those networks takes place only with technocratic NGOs, not with other types of NGOs. Furthermore, the failing outcomes of government organized NGOs in promoting sustainable production are in line with the views articulated in the discourses regarding the Peruvian government. What all discourses agree on is that government has to lead the regulation of polluting SMEs.

Low institutionalization of sustainable production in governmental decision making has been highlighted in previous chapters. Governmental economic policy prioritizes what is in line with global market demands such as the export of goods. To do so, the central government has established policies and an institutional framework for companies that export goods. Prime ministers play active roles in connecting actors in Peru to global market actors, establishing favorable conditions in the country for foreign capital investments and prioritizing the support of larger scale domestic and foreign companies. Authors such as Jimenez (2001) and Zunkel (2005) have highlighted this characteristic of the liberal model of Peru: the exportation of primary goods is prioritized, while local industrial development is hardly cared for. Central government views regarding SMEs focus on development, modernization and competitiveness. However, national policies that support sustainable production are hardly operationalized. For instance, according to Felipe-Morales (Revista Agraria, 2000) and Seegers (2008, interview) governmental policies supporting organic agriculture are missing. According to the proponents of the business upgrading discourse and the market democratization discourse the Peruvian government does not show commitment neither with sustainable SMEs nor with the development of local sustainable markets, for instance, in investing and providing incentives to SMEs that show good environmental and social practices. However, a more positive sign of governmental interest has been the launching in 2008 of the award 'Premio Presidente a las MYPE' (now named 'Premio Nacional a la MYPE') to award SMEs' best practices and the establishment in 2010 of the funding scheme Innovate Peru FIDECOM for technological research in SMEs. Although governmental initiatives, including INCAGRO, testify of at least some interest in SMEs development, the performance of SMEs in terms of environmental sustainability in the three domains under study is not central for the Peruvian government (Crece MYPE website, 2012; INCAGRO website, 2011; Innovate Peru website, 2011).

The lack of governmental attention to re-enforcing links between economic growth and environmental protection has been emphasized by the proponents of the cleaner production discourse and the business upgrading discourse. Environmental protection is left to market instruments (e.g. certification schemes) and non-state actor interventions (e.g. companies and NGOs). Proponents of all discourses perceive that economic rationalities are at the core of the governmental policy decision making and ecological rationality is only valid if it contributes to economic development. In the words of the former (2006-2011) president of Peru Alan Garcia, nature (forest, land, mining, oil and sea) is a resource that has to become input of the national production system in order to generate employment and progress in the country (Garcia, 2007).

8.5.2 International cooperation agencies

Under the label of international cooperation agencies several types of organizations are clustered, including international NGOs, (inter)governmental agencies, research centers and consultancy firms. International cooperation agencies are connected to local NGOs through supranational and national platforms and/or by direct connection.

International cooperation agencies play essential roles in the networks of sustainability of SMEs. In addition to funding as a main support strategy, international cooperation agencies build capacities of local NGOs and SMEs by providing them with knowledge to improve production and commercialization of SMEs (Engelhardt, 2008, interview). Moreover, these agencies support lobbying at the international level, facilitating matchmaking of SMEs with market actors and supporting platforms of NGOs and SMEs (Mechielsen, 2008, interview; Verhallen, 2008, interview). Particularly, the matchmaking role is becoming more important as globalization might provide potential benefits for SMEs, for instance, matching developing countries' small producers with developed countries' large retailers (Verhallen, 2008, interview). International cooperation agencies are linked to a large number of local NGOs, often in more than one of the domains investigated. For instance, ICCO is linked to RAE Peru, Peru 2021 and CEDAL; OXFAM-NOVIB is linked to Grupo Ecological Peru and CEDAL but HIVOS only to ANPE. Cooperation between international cooperation agencies and local NGOs is based on their common views regarding priorities and strategies of intervention as far as SMEs are concerned. However, as was established in Section 8.3, the sharing of views between networks does not automatically ensure cooperation.

International cooperation agencies are experiencing changes in their intervention policies. Compared with the 1980s and 1990s when the main focus was on rural development, the current financial support of international cooperation agencies is based on market rationality (Engelhardt, 2008, interview; Seegers, 2010, interview). In the last 20 years, for instance, the agenda of international NGOs has been shifting from social-centered to economic-centered (Verhallen, 2008, interview) and attention is increasingly put on directly funding SMEs (Langbroek, 2008, interview). To date, their main focus is on strengthening value chains, setting business development services and market intermediaries for SMEs to access national and international markets, particularly putting emphasis on sustainable production and market (Engelhardt, 2008, interview; Seegers, 2008, interview). Increasingly the funding of projects of local NGOs is based on business plans, and local NGOs are obliged to submit reports periodically to their funders. Agencies such as COSSUDE and EU are in the forefront of these changes in their funding schemes but organizations such as ICCO still keep quite flexible funding criteria (Alvarado, 2009, interview). This pressure for professionalization forces local NGOs to become more sophisticated organizations by hiring managers and implementing management tools to handle projects (Alvarado, 2009, interview). Projects of local NGOs are evaluated by international cooperation agencies by quantitative measuring and short-term results. As Alvarado (2009, interview), president of RAE Peru, states: 'changes in international cooperation agencies contribute to achieving project targets and improving organizational planning, but they weaken our capacity of political influence, social activism and reflection. The risk is that NGOs become just like consultancies and project operator organizations'. The major emphasis in funding criteria on quantitative achievements and on further involvement of NGOs in the market brings organizational tensions, either towards further specialization or towards splitting of the organization (Verhallen, 2008, interview). Changes are mainly due to the pressure of developed countries national governments to increase efficiency in funds allocation for development cooperation. For instance, Dutch NGOs were expecting cuts in their governmental funding for development cooperation in 2010. As a result, international

NGOs are changing their cooperation strategies concentrating in particular regions and issues. For instance, international NGOs are retreating from Peru, as for instance CORDAID intended to do in 2012 and OXFAM-NOVIB did in 2011. HIVOS and ICCO are reformulating their intervention strategies (Seegers, 2010, interview). Moreover, international cooperation agencies are looking at the government as a new channel for their development assistance. For instance, the governmental funding schemes and awards initiatives for SMEs (e.g. Innovate Peru, 'Premio Nacional a la MYPE' and INCAGRO) are financially supported by international financial institutions (e.g. World Bank), international cooperation agencies (i.e Belgian Technical Cooperation) and companies (e.g. British Tobacco) (Crece MYPE website, 2012; INCAGRO website, 2011; Innovate Peru website, 2011).

The relationships between international cooperation agencies have particularities in each of the three domains. In the case of the agencies engaged in the networks of organic production, most of the agencies are international NGOs. International NGOs such as OXFAM-NOVIB, ICCO and HIVOS closely work with local NGOs and are part of international platforms that promote sustainable SMEs worldwide. International NGOs provide funding and connection with markets but the self-organizing capacity of local NGOs make possible intervention. In the case of business social responsibility networks, the range of international cooperation agencies is much wider. For instance, agencies can be larger business foundations, international NGOs, and international financial institutions. International agencies are connected to local NGOs through regional platforms or by direct connection. While some international agencies (e.g. AVINA) are only connected to one of the two business social responsibility networks identified, others (e.g. ICCO) are connected to both networks. But, connection in this domain is limited to actors with common views. Local NGOs are very proactive and collaborate closely with their funders. Finally, in the case of sustainable production networks, they involve wider types of international cooperation agencies, for instance, international NGOs, inter-governmental agencies, international financial institutions, governmental donor agencies and research institutions. International cooperation agencies have taken the leadership in promoting cleaner production and appropriate technology. Local NGOs and government agencies, in this domain, are mostly project operators and their self-organizing capacity is low.

8.6 Reflecting on theory

Theoretical reflection on the research outcomes is presented against the backdrop of current debates on social network theory and ecological modernization theory.

8.6.1 Social network theory

This section reflects on three key concepts addressed in the social network theory; configuration, cooperation and power relationships, on the basis of the findings presented in previous chapters.

Configuration

Castells (2004) in his network society theory distinguishes between the space of places and the space of flows. In the context of this thesis the space of flows is the dimension where actors promoting sustainability of SMEs exchange resources at long distance mediated by information and communication tools, and the space of places comprises actors promoting sustainability of SMEs exchanging resources in the local situation. The networks and discourses of sustainability of SMEs are part of both types of spaces. Although actors of the networks of sustainability of SMEs are rooted in their local realities, their connections are not limited to the physical contiguity but go beyond geographical boundaries (Figure 5.1, Figure 6.1 and Figure 7.1). Along with networks, discourses might also flow beyond geographical boundaries and they can be endorsed by actors located in different part of the world. These dynamics help to understand better, at a theoretical level, the relationships between networks and discourses and the relationships between the space of flows and the space of places.

Following Castells' theory, networks of sustainability of SMEs encompass connections between their affiliates inside and outside of the research domains. Additionally, what the research has shown is that the structuring of networks is not only made up of actors and their connections but also of platforms. Castells' theory does not highlight platforms as a vital structure within networks. A platform is a coordination point of actors, usually led by a node, with a particular goal within a network or between networks. Platforms can be temporary or long-term, and of local, regional or global reach. For instance, in the sustainable production networks five platforms have been identified at the regional level in Latin America: the Latin American Cleaner Production Platform, the CAB Regional Platform of Cleaner Production and the technological transfer platform, RESIRDES and CYTED. In the business social responsibility networks two platforms have been identified at the regional level in Latin America: the Red Puentes International (this platform has also its local equivalent in Peru) and the Forum Empresa. In the organic production networks two platforms have been identified at the regional level in Latin America: MAELA and GALCI; and two at national level in Peru: CONAPO and the 'Peru, Country Free of Transgenics' platform. As the densities of connections, coordination and exchange of resources are greater in the platforms than in their single actor members, these platforms determine to a fair extent the capacity of networks to survive and expand. For instance, the networks of the organic production domain are stronger than the networks of sustainable production domain due to the connection and cohesion roles of their platforms in terms of, for instance, involving a large number of NGOs and SMEs.

Discourses of sustainability of SMEs emerge from local actors in the networks and once they are launched in the space of flows, they are endorsed by other actors located anywhere in the world. Just to give an example: subscribers and critics of the market justice discourses can be found in far away countries such as the Netherlands and US, not just in Peru; similarly for the sustainable market discourses. Although discourses of sustainability of SMEs are part of the space of flows, they are embodied in the networks. Networks are not just carriers of discourses, but the space where new discourses are generated. The discourses of sustainability of SMEs have emerged from their mother discourses; the liberal market discourse and the social movement discourse under the framework of networks. Furthermore, discourses are the source from where goals and plans are structured in the networks in terms of what is, and how sustainable SMEs should be, stimulated for more environmental friendly production. For instance, the goals and plans of the Red Puentes platform of the social justice network are rooted in the market justice discourses and the Forum Empresa platform of the business network is founded in the sustainable market discourses. Therefore, the discourses of sustainability of SMEs are internationally extended and have incorporated elements of older discourses under the framework of network relationships.

Tensions in the networks and discourses of sustainability of SMEs give evidence of the struggling within the space of flows. A major tension for these networks is that of uniformity and integration at the global level versus diversification and fragmentation at the local level. This phenomenon has been observed in all networks identified. For instance, regional and global promoters of sustainable production such as UNIDO, SECO, GTZ, Practical Action, UNEP, OAS, AECI, DED, IDB, EU and CORDAID have more connections than the local actors in Peru such as CER, ITACAB, CONCYTEC, Practical Action, CITEs and IPES. Moreover, local actors in the networks have more connections with their regional and global 'partners' than among them. So exchanges and coordination in the flows are stronger at the global side of the networks, though most of operative roles towards SMEs take place at the local level. These characteristics show that networks are in permanent tension between opposing forces along their structure. Discourses are not free of tensions as well. A major tension among the discourses of sustainability of SMEs is due to their position between adversarial poles; between the 'communitarian' and 'individualistic' poles and between liberal market and social movement poles. Tensions between discourses are unavoidable as each emphasizes particular values. While the market access discourse, the corporate responsibility discourse and the cleaner production discourse emphasize 'liberal market' and 'individualistic' values, the democratization discourse, the appropriate technology discourse and the upgrading discourse emphasize 'self-sustaining', 'rights' and 'communitarian' values. In addition, the market adaptation discourse emphasizes 'communitarian' and 'liberal market' values. However, the emphasis in certain values does not mean discourses deny liberal market as it has embedded in all discourses. Although all adversarial poles condition tension between discourses, at the core of the tension is the question whether individualistic or communitarian values are central to organize economic relationships. Furthermore, the polarization of discourse perspectives either to the liberal market corner or to the social movement corner pressure SMEs to business as usual and stagnation rather than stimulate them towards sustainability. This is so because polarization undermines cooperation and enhances mistrust of social and market actors. As a result, a fertile ground for promoting sustainability of SMEs cannot be built, especially given the fact that the national government is lagging behind in promoting sustainability in SMEs. Finally, tensions also are found in the relationship between networks and discourses of sustainability of SMEs. While networks are dynamically establishing new connections and engaging new actors, discourses tend to keep networks static. This tension between expansion and contraction makes networks display particular patterns in endorsing discourses. Networks display three patterns of change in configurations; 'unidirectional shift', 'converging shift' and 'diverging shift'. Each pattern applies for one of the research domains. Thus, we see a 'unidirectional shift' for the organic production networks, a 'converging shift' for the business social responsibility networks and a 'diverging shift' for the sustainable production networks.

Cooperation

Discourses do influence cooperation. What stimulate cooperation are the communalities of the discourses in terms of the common interest of agents to support sustainable SMEs. NGOs have established cooperation with other agents (e.g. national governmental agencies, SME associations, larger companies and municipalities) that share their views on how to encourage sustainable SMEs. The research shows that cooperation happens despite differences and conflicts in the networks, the organic production domain being the best example. Networks with similar discourses, however, do not necessarily work in cooperation. This was especially observed in the sustainable production domain. Lack of connections among actors and self-interest are the conditional factors for this scarce cooperation. Nevertheless, the discourses of sustainability of SMEs complement each other in their views to cope with the environmental improvement of SMEs rather than that they clash. A single discourse can not capture the whole heterogeneity of NGO perspectives on how to achieve such improvement. This complementary character of the discourses establishes conditions for further cooperation.

Cooperation is not uniform among all actors. Cooperation among actors affiliated to the same discourse is more feasible than among actors affiliated to different discourses. This irregular cooperation has to do with the type of values actors endorse; being either individualistic or communitarian. Currently individualistic values embodied in the liberal market discourse are central in SMEs economic relationships. However, both 'communitarian' and 'individualistic' values are needed to improve sustainable practices in SMEs. 'Communitarian' values enhance the mobilization and cohesion of SMEs to perform successfully in sustainable markets and individualistic values strength individual entrepreneurship and competitiveness to develop and access to sustainable markets.

The fact that discourses of sustainability of SMEs do not share core viewpoints with social movement and liberal market discourses, also contributes to establish channels of cooperation among actors. Yet, strands of those adversarial discourses are to some extent mixed up in the discourses of sustainability of SMEs, causing a reasonable tension among network actors. However, the adversarial core views have been overcome. As a result of this process of re-arrangement of usual conflicting perspectives in a new set of discourses out of adversarial ideologies, the discourses of sustainability of SMEs embody new storylines that have to do with the call for a 'just' engagement of SMEs in profitable local and global value chains. I consider this process of formation of new discourses from previous opposing discourses as a form of 'reflexivity'. This reflexivity of the discourses is a result of the nature of networks, such as their capacity of self-expansion and the possibility of recombining information and adaptation (e.g. Castells, 2004: 9). Reflexivity of discourses favors cooperation of the networks promoting sustainability of SMEs since actors are not tied to the aforementioned opposing discourses. As a result, actors in the networks of sustainability of SMEs are more plausible to come together and cooperate. Cooperation will increase if dialogue and values of 'reflexive' thinking become central in supporting sustainable SMEs. Cooperation will be undermined in the networks if imposition of an 'orthodox capitalism' or a radical 'against capitalism' perspective is embraced by influential agents.

In addition to discourses, the success of cooperation is influenced by other factors such as globalization of liberal market economy, availability of funding and rewarding of legitimacy and prestige.

Power relationships

Power relationships in the networks promoting sustainability of SMEs are characterized by asymmetry. This asymmetry has to do with the performing of particular tasks by clusters of actors in the network. For a better explanation of this statement I will use the metaphors 'network body', 'network mind', 'network hands' and 'network target'. The 'network body' is the structure of the network (e.g. platforms, connections, and information and communication technologies). The 'network mind' constitutes the network actors (e.g. international NGOs, foundations and donors) that govern the network, establishing policy guidance and providing funding. The 'network hands' are the network actors (e.g. local NGOs) that put in practice the policy guidance by implementing or operating projects with SMEs. The agency of 'network hands' is limited to the operationalization of the guiding policies of 'network mind' actors and does not include elaborate the rationality behind of the interventions on SMEs. The 'network target' is constituted by SMEs. SMEs are considered 'powerless' by 'network mind' and 'network hands'. Thus, power relationships between 'network mind', 'network hands' and 'network objective' are asymmetrical depending on the capability of actors to establish policy guidance and providing funding. It means that certain actors in the networks have the ability to provide direction, supervision and control, based on knowledge and funding. This asymmetry can be strong or weak, depending of the degree of control or emancipation. For instance, in the governance of the eco-efficiency network UNIDO/UNDP monopolizes power, CER, the local NGO, implements the global guidelines of UNIDO/UNDP, and SMEs are just passive beneficiaries. But in the ecological farming network SMEs are organized in ANPE, the local NGO, and as representative of their affiliated SMEs ANPE coordinates directly with HIVOS, ADG or IFOAM to implement projects.

   Power relationships in the networks of sustainability of SMEs do not follow the typical binary logic of resistance/domination or inclusion/exclusion between the space of flows and space of places, as argued by Castells (2004). Power is distributed over both spaces and any actor can exercise it to command or emancipate from other actors in the networks. SMEs hold power, as international NGOs do. SMEs are not as powerless as thought by the 'network mind'. SMEs in the networks have the capability to legitimate or challenge the power of international cooperation agencies (e.g. OXFAM-NOVIB, ICCO) and local NGOs (e.g. Grupo Ecologica Peru). Based on the findings of this research, I argue that SMEs have power but of another nature than the power of the aforementioned 'network mind' actors. I call it 'anchoring' power. I define anchoring power as the capability of actors to be anchored in local social webs of cooperation, reciprocity and exchanges of goods and services, and representativeness in the networks. 'Anchoring' power is a key resource of actors (e.g. SMEs, local NGOs) to gain position in the networks and bargain and negotiate with other network actors (e.g. international cooperation agencies and local NGOs). The power of international cooperation agencies is founded on the capability to organize knowledge and information and distribute them along the network with the help of information and communication tools. I call this 'informational' power. So, actors in networks of sustainability of SMEs are in a power struggle, each of them using their 'anchoring' or 'informational' power as weapon.

   Relating the concepts of informational power and anchoring power with the network society theory, it can be said that, with regard to the subject of this thesis, power in the space of flows is based on information and access, and the spreading out of uniformity and integration. But in the space of places it is based on local anchoring, diversity and physical contiguity. SMEs are gaining power in networks as they are getting connected to the space of flows but keeping their roots in the space of places. The emerging of new NGOs representing small scale enterprises (e.g. ANPE, CEPICAFE, APROMALPI, CEPIBO and REPEBAM) and NGOs affiliating with small scale enterprises (e.g. Grupo Ecologica Peru) along the country, and their capability to assemble their views in a set of new discourses shows that anchoring power constitutes an advantage of SMEs not only to gain a position in networks in times of globalization but to become a source of 'fresh' views beyond the dichotomy of liberal market and social movement. In contrast to what Castells states in his theory of space of flows (e.g. Castells, 2004: 22, 23, 36), this investigation does not show that the space of places is by necessity subordinated to the space of flows. What is observed in the networks of sustainability of SMEs is that global actors intend to control local actors and local actors try to influence global actors. For instance, while local SMEs show power by refusing to cooperate, by legitimating interventions of NGOs or by requesting accountability to NGOs, international NGOs show power by deciding which local NGO or SMEs is subject of funding. The rise of the network society provides a new floor for local and global actors to interact with the help of information and communication tools that connect them at a global level. In engaging with global networks, local NGOs and SMEs are getting included in the space of flows. However, there is no dominant actor who decides which actor will be included or excluded in the network, as Castells claims. Neither global nor local actors are all-dominant in the networks of sustainability of SMEs; none of the two is in control of the discourses. Mutual interests or needs bring local NGOs, SMEs and international NGOs to connect to each other in the networks of sustainability of SMEs. Furthermore, it is not only the global that overwhelms the local, but also the local that is diffusing and spreading out globally. Local NGOs and SMEs take advantage of this new social structure moving from being a 'target' of global actors to gaining position as an actor in global networks. For instance, organic producer organizations, including ANPE, are organized at the global level through the Intercontinental Network of Organic Farmers Organizations (INOFO) that coordinates the participation of worldwide producers at IFOAM. This means that although the space of flows is a new dimension, it is influenced by places since SMEs and local NGOs, the main actors of the networks of sustainability of SMEs, mostly operate at the local level. Thus, power is distributed over both the space of places and the space of flows.

8.6.2 Ecological modernization theory

Many of the changes described appear to be in line with the tenets of ecological modernization theory. This section is dedicated to explore if this indeed is so, for three major aspects: institutional transformation, appreciation of the liberal market in environmental operations, and environmental rationality.

NGOs promoting sustainability of SMEs have overcome the traditional position of NGOs against the liberal market. They are undergoing institutional changes in enhancing cooperation with other actors (e.g. SMEs and larger companies), connecting with local and global actors in networks of global reach, and standardizing institutional operation to improve the position of SMEs in sustainable production and markets. NGOs have become the main actor in promoting sustainability within SMEs by promoting eco-efficiency measures, organic certification and social and environmental standards. The fact that NGOs take the position of market agents to promote sustainability of SMEs is a variation of ecological modernization. Under the theory of ecological modernization companies usually take this position, not NGOs. Although companies are much better equipped to operate effectively and efficiently on markets than NGOs, they lack capacities to connect directly with most SMEs. For instance, companies alone are not able to organize the dispersed and varied SMEs to include them in their value chains. Companies, especially foreign ones, and SMEs, particularly, micro and small scale enterprises, have social and cultural gaps that matters to engage them in business endeavors. Therefore, companies need to establish or cooperate with NGOs or similar organizations to overcome such gaps and reach most SMEs. In the investigated cases, it appears to be easier for NGOs to gain capacities to operate on markets than it is for companies to gain capacities to understand the reality of most SMEs. Furthermore, the role of NGOs as market agents has come as a result of the demands of market actors (e.g. buyers, exporters) for building ties with SME. According to the research findings, NGOs will keep gaining their position as market agents for SMEs as long as they adjust their managerial and operational structure to do so. So it can be argued that the capacity of NGOs, especially of local NGOs, does not primarily come from high capital resources or access to global resources and markets but from the capacity to establish cooperation ties with other NGOs, SMEs and to be anchored in local social networks. This emergence of NGOs as market agents means the change from 'typical' NGOs to 'hybrid' NGOs: consulting or company-like NGOs. As market agents, NGOs play not only civil society roles but also business roles. However, NGOs performing as market agents are still different from companies. Typical NGOs, as founded in 1980s, will become marginal and new types of hybrid NGOs will become more central in supporting sustainability of SMEs. This hybridization of NGOs supporting SMEs is expressed in changes in organization, strategy and discourse; for instance, implementing a more standardized work style and introducing business tools within their organization and in their SMEs clients, affiliating SMEs within their organizational structure, accepting, or even, advocating liberal market perspectives and cooperating in NGO networks of global scope. The reasons for this central role of NGOs and the change to a more hybrid profile have to do with the situation of the Peruvian state, the market actors and SMEs in Peru. The Peruvian state does not take leadership in facing environmental degradation related to SMEs, SMEs institutions are fragmented and market actors (e.g. companies, banks, consulting) are lagging behind in promoting sustainability of SMEs. To be precise, in their efforts to support SMEs the Peruvian state and the market actors prioritize economic considerations, but marginalize environmental concerns. NGOs fill the gap of an ineffective Peruvian state and short-falling market actors in improving the environmental performance of SMEs.

Discourses of sustainability of SMEs give equal weight to the need of further economic and ecological performance of SMEs. The discourses endorse the introduction of environmental rationality in management and operations to measure environmental impacts, optimize and monitoring performance, and report and rate SMEs on their environmental performance. However, the discourses emphasize different methods. The business upgrading discourse, the corporate responsibility discourse, the cleaner production discourse and the market access discourse emphasize global standards (e.g. ISO 14001, organic certification and ethical trading initiative) and the market adaptation discourse, the market democratization discourse and the appropriate technology discourse emphasize methods more tailored to the specific situation of SMEs (e.g. eco-efficiency measures, appropriate technologies). This means that although all discourses of sustainability of SMEs value the importance of technology and innovation to make SMEs more sustainable, the last group of discourses widens the scope of methods for ecological improvement. So, ecological rationality in the discourses of sustainability of SMEs implies not only the most sophisticated methods and technologies for ecological improvement of SMEs but also straightforward methods that suit better to the local and social context of SMEs. Ecological modernization theory usually emphasizes the use of environmental standards (e.g. labels, ISO standards, life cycle analysis) and the use of renewable energies for optimizing environmental performance. But it does not much elaborate on methods integrating social and environmental concerns, for instance: social responsibility standards, organic certification and participatory guaranty systems. Many of the most advanced methods to improve environmental performance hold limitations for SMEs in Peru as most of them are small and micro enterprises. Methods integrating social and environmental concerns underlined in the discourses of organic production and discourses of business social responsibility contribute also to ecological modernization of SMEs since social and environmental concerns are strongly interlinked in SMEs. Similar to the mutually reinforcing role of economic and ecological concerns for larger companies, social and ecological concerns play reinforcing roles for SMEs. Therefore, it can be established that while the sustainable market discourses match with the ecological modernization discourse, the market justice discourses goes beyond it. A difference of the sustainable market discourses that highlight the market advantage for SMEs of improving environmental performance, the market justice discourses highlight the values of social justice in mediating the access of SMEs to markets. Additionally, two market justice discourses, the market democratization discourse and the market adaptation discourse underline the values of harmony with nature instead of approaching nature as resource. However, the discourses of sustainability of SMEs and the ecological modernization discourse share views on the need to improve both economic and ecological performance of SMEs to access sustainable markets. So, the findings of this study suggest the need for a broader understanding of ecological rationality, which include the use of sophisticated and easy-to-apply environment centered and social-environmental integrated methods, and different approaches on market and nature to improve the position of SMEs in sustainable production and markets outlined in the discourses of sustainability of SMEs. This broader understanding contributes to the raise of a 'reflexive' ecological modernization discourse to bridge the concerns of sustainability and fair market access of SMEs in Peru.

8.6.3 Future pathways for environmental governance of SMEs

We end this chapter with a brief look into the future of environmental governance of SMEs.

The research shows that the networks of sustainability of SMEs are the foundation of an emerging governance mechanism of SMEs. Within those networks, NGOs and SMEs constitute the main actors in planning, channeling resources (e.g. funds, knowledge) and implementing projects in SMEs to improve their environmental performance *vis-à-vis* their economic concerns. Governmental agencies are virtually absent in the networks studied, or play only secondary roles. Larger companies also play secondary roles in the environmental governance structure of SMEs. In the case of a more central role played by governmental agencies it has usually been a failure, for instance, in the sustainable production networks. NGOs have overtaken the promoting roles of the Peruvian state more effectively. The failing role of the Peruvian state in handling environmental challenges of SMEs provides evidence that the state-driven governance model cannot bring solutions to pressing environmental challenges related with industrial development and market expansion in Peru. The crisis is of legitimacy and governance model. Different from a state-driven governance that is centralized and mandatory in establishing decision making and policies, network governance is decentralized and discretionary. Decision making and policies are not established by a privileged unique actor or cluster of actors, but by nodes of coordination at local, regional and global levels, often in the form of platforms. Discourses, cooperation and power struggles are forces that shape the environmental governance of SMEs in the networks. However, it has also been observed that coordination among platforms of different networks, or even within a network itself, does not always go easily. This has to do with the diverging emphasis of the discourses, either on social justice or on environmental sustainability. As a result, cooperation is sometimes limited to certain networks or clusters within a network. Moreover, actors are involved in struggles for power positions. According to our research, struggle strengthens the governance structure rather than weakening it, as we observed regarding the emancipation of SMEs in the organic production domain. The networks of sustainability of SMEs show that NGOs and SMEs can play central roles in governing networks without a significant input of larger companies and the national government.

Governance of SMEs has to be based on environmental considerations to attain the sustainability of SMEs in the long-term. This can be achieved by establishing environmental policies based on sustainable business models that take into account the heterogeneity of SMEs in Peru. Two main pathways of environmental governance of SMEs are drawn from the case studies. The first pathway is emphasized by the sustainable market discourses and the second pathway by the market justice discourses. The first one, calling for market access, consists of further engagement of SMEs with sustainable global markets. It means further ecologically sound modernization of SMEs oriented to international markets, further adoption of cleaner technologies and further clustering of SMEs in larger business associations and cooperatives. This process implies the integration of SMEs in global sustainable markets as suppliers. The second pathway, calling for market development, consists of furthering the setting up of sustainable SMEs and the engagement of SMEs with local sustainable supply chains in a diversified, adapted and decentralized way throughout the country. This means that SMEs might remain small in size but increase in production scale by stimulating the formation of associations, cooperatives, conglomerates and clusters at different ecological niches, further use of low-cost technologies and decentralization of production aimed at developing local sustainable markets.

As market agents, NGOs connect both pathways by supporting SMEs to access global sustainable markets and develop local sustainable supply chains. This inclusive role of NGOs put them at the forefront for environmental sustainability of SMEs, in comparison with larger companies and the Peruvian state, which, at best, focus on the first pathway. In emphasizing only one of the pathways, those actors are not taking into account the heterogeneity of SMEs in capacities, interests and suitability. Furthermore, the anchoring power of SMEs might be an advantage of the emerging governance scheme, as it triggers sustainability at local level, for instance by involving sub-national and local governments. Global sustainable markets contribute to the development of local sustainable supply chains, and vice verse as the case studies show. Global and local sustainable markets are not conflicting but reinforcing mechanisms for improving the environmental performance of SMEs in Peru.

The networks constitute new market relations beyond the typical triangle: state, civil society and business. In the networks, not a single actor – including NGOs – stand for environmental improvement of SMEs, but the entire network. Sustainability and fair access of SMEs to markets are the binding perspectives that tie all actors in the networks. According to the networks analyzed, three models of environmental governance are outlined. First, there is a highly effective model represented by the networks of the organic production domain. Second, a medium-effective model can be identified, represented by the networks of the business social responsibility domain. And third, a low-effective model exists, represented by the networks of the sustainable production domain. The degree of effectiveness is related with the self-organizing capabilities of NGOs and SMEs to lead processes of environmental improvement in SMEs. SMEs tied to the networks of sustainable production and of business social responsibility show higher levels of fragmentation, isolation and dependency than SMEs tied to the networks of organic production. These characteristics of SMEs affect the permanency of environmental changes in SMEs. Thus, in the light of this research it can be stated that the networks of sustainability of SMEs provide both an analytical tool and useful models for environmental governance and thus constitute a good example of a new environmental governance mechanism: governance without the state.

Chapter 9.
Conclusions and recommendations

In the previous chapters, the changing roles of NGOs in promoting sustainability of SMEs in Peru were investigated through the perspectives of networks and discourses. Three cases were analyzed: organic production, business social responsibility and sustainable production. This concluding chapter starts with briefly re-introducing the background, problem statement and research questions (Section 9.1). Subsequently, the chapter summarizes the most significant conclusions from the three cases on the kind of networks and discourses, as well as on the major changes in networks and discourses (Section 9.2 and 9.3). Section 9.4 analyzes to what extent these findings can be generalized outside Peruvian SMEs. After, discussing the implications of these findings for theoretical debates on network society theory and ecological modernization theory (Section 9.5) recommendations for policy and research are formulated (Section 9.6).

9.1 Introduction

The importance of small and medium-sized enterprises (SMEs) in terms of employment and income generation has been recognized worldwide. In the region of Latin America, SMEs are responsible for 25% and 40% of employment, and between 15% and 25% of goods and services. In Peru, SMEs are responsible of 85% of the employment at the national level and they represent 98% of the total companies registered. Around 12% of SMEs, organized in associations, clusters, and cooperatives or as single companies, are dedicated to productive actives, the others are engaged in commercial and services activities.

However, the growth of SMEs is also responsible for significant depletion of natural resources, environmental degradation and threats to human health. Particularly SMEs in production sectors that work with transformation of materials and use of energy, such as textile, leather, food and metal industries, often pose risks to workers' health and community's environment. Also SMEs producing and commercializing food contribute to environmental problems, as the intensification of agriculture (with the use of pesticides and artificial fertilizers) causes impacts on soil and water and on human health. Environmental pollution by SMEs often goes together with low quality employment in terms of productivity, income and working conditions. Environmental degradation represents a cost for Peru of about 4% of GDP due to increased morbidity and mortality and decreased productivity. So the change of polluting SMEs to sustainable SMEs is an urgent need in Peru, to reduce the environmental burden, and increase their competitiveness and job quality.

Next to command-and-control strategies, the national government has been encouraging industries to voluntarily adopt environmental standards, cleaner technologies and energy saving measures. However, the priority of the government has been to support larger companies and SMEs attached to international markets. The national government shows institutional shortcomings in addressing polluting SMEs, especially those that are not linked to international markets. Hence, following this state failure, the role of non-governmental actors in promoting and enabling environmental friendly practices of the industrial sector has received increasing attention, not only in Peru but worldwide. Especially in developing countries NGOs are closer to SMEs than state organizations, and they are increasingly taking over tasks and duties of the state in the social and economic development of SMEs.

NGOs support SMEs by providing them with resources, knowledge and technology, by encouraging them to develop co-operative relationships and by creating the necessary conditions to improve their environmental performance. NGOs play an important role as intermediary organizations in providing links between the business sector, development cooperation agencies and the government. Particularly, NGOs are important in encouraging the development of sustainable SMEs, by connecting global and local actors claiming to be the natural vehicles of development and local empowerment. Recently, NGOs engaged in global networks are adopting discourses of sustainability to promote sustainable SMEs. The most relevant networks and discourses in terms of promoting a balance between environmental protection and economical growth in SMEs are those involved in organic production, business social responsibility and sustainable production. These developments mean that NGOs are changing their traditional role as they become more embedded in supporting productive activities of SMEs for local and global markets. However, little is known about the roles, impacts and successes of NGOs in stimulating the transition of SMEs towards sustainability, especially in developing countries.

To shed more light on the new roles of NGOs in promoting sustainability of SMEs in Peru, this thesis explored the actual and potential changes in networks and discourses that NGOs supporting SMEs are undergoing. It uses the domains of organic production, business social responsibility and sustainable production as empirical cases, and ecological modernization theory and network society theory as theoretical lenses. The thesis aims to understand how networks of NGOs support SME sustainability in Peru, and which discourses these networks apply in supporting SME sustainability. Hence, the following central research aim is established: provide a better understanding of the changing roles of NGOs in promoting sustainability of SMEs in Peru using the perspectives of networks and discourses. Three research questions have been outlined for this research: first, what are the networks of NGOs promoting sustainability of SMEs involved in the domains of organic production, business social responsibility and sustainable production in Peru, and what are the main changes in time in these networks? Second, what are the main discourses fostering sustainability that prevail and are articulated in these networks of NGOs and what are the main changes in time in these discourses? And finally, how to understand and assess the actual, new and potential roles of NGOs in promoting sustainability of SMEs in terms of network society theory and ecological modernization theory?

9.2 General characteristics of networks and discourses

This section answers the first and second research questions by focusing on the main conclusions in terms of the general characteristics of the networks and the discourses of NGOs involved in the domains of organic production, business social responsibility and sustainable production in Peru. The main conclusions are organized in two themes.

9.2.1 Diversification of networks and emerging of new types of NGOs

The networks involved in promoting the sustainability of SMEs, as identified in this research, are the following: the agro-ecological network, the organic market network and the ecological farming network in the first case study (organic production); the social justice network and the business network in the second case study (business social responsibility); and the eco-efficiency network, the appropriate technology network, the cleaner technology network, the technological innovation network and the urban cleaner production network in the third case study (sustainable production). These networks are structured in interlinked platforms operating at local, national, Latin American and global level. Platforms are a vital structural component within the networks and between promoting sustainability of SMEs. Although platforms might involve SMEs, larger companies and government representatives, the key actors coordinating and channeling resources in the platforms are usually NGOs. The networks promoting sustainability of SMEs are composed by diverse types of NGOs. Next to conventional NGOs as key actors, producer NGOs, market NGOs, business NGOs, technocratic NGOs and government organized NGOs (GONGOs) have emerged in the networks. This means that the networks currently have NGOs run by SME entrepreneurs, businessman, engineers and governmental officers. In line with more network diversity in the organic production and sustainable production domains, more diversity of NGOs is found in these networks. GONGOs are governmental agencies performing as key actors in the sustainable production networks. GONGOs operate as typical NGOs and work closely with technocratic NGOs. The common characteristic of those new types of NGOs is that they are more market-driven organizations than the conventional NGOs.

As a result of this diversification of networks and NGOs the struggle for leading positions in the network platforms and the competition for scarce funding and for operation of projects of international cooperation agencies have also intensified. This diversification of NGOs and, above all, the increase of market-driven NGOs occurs in response to the business growth and market demands of SMEs. In line with findings reported by Bebbington and Thiele (1993), Diaz-Albertini (2001), Beloe *et al.* (2003) and Sanborn and Portocarrero (2008), this study evidences that after the 1990s NGOs in Latin America are becoming professional organizations, and are departing from philanthropy, charity and activism against the state and the economy. However, networks and NGOs promoting sustainability of SMEs do not totally abandon former perspectives, as their focus on solidarity illustrates (see Section 9.3).

9.2.2 Discourses emphasize either market justice or sustainable market

The discourses identified in the networks of NGOs promoting sustainability of SMEs are the following: market adaptation, market access or market democratization in the domain of organic production; business upgrading and corporate responsibility in the domain of business social responsibility; and cleaner production and appropriate technology in the domain of sustainable production. These seven discourses can roughly be divided in two groups, one emphasizing market justice and the other emphasizing sustainable market. The two groups are different in their interpretation of environmental reform and sustainability. In the market justice discourses emphasis is put on inclusion of emerging new economic actors such as SMEs and environment is seen as something to be protected, defended and stood up for. In the sustainable market discourses emphasis is put on successful market performance, and environment is seen as business opportunity and a factor contributing to a company's legitimacy ('its license to operate'). The market justice discourses and sustainable market discourses are also different in their position toward social movements and the market. The origins of these discourses lay in two long-standing, antagonist, so-called 'mother' discourses: the liberal market discourse on one hand and the social movement discourse on the other hand. While the cleaner production discourse and the corporate responsibility discourse have their origins in the liberal market discourse, the market democratization discourse, the market adaptation discourse, the market access discourse, the market adaptation discourse, the business upgrading discourse and the appropriate technology discourse have their origins in the social movement discourse. Of the latter discourses, the market access discourse has most strongly moved away from its origin. The market access discourse has strongly been influenced by the liberal market discourse as it widely diverges in their storyline from the other two organic production discourses. Besides the emphasis on either market justice or sustainable market, the discourses promoting sustainability of SMEs are a new assembling of views about production, the environment, the market and globalization applied to the reality of SMEs, beyond the conventional discourses of liberal market and social movement. In this way, under the notion of sustainability long-standing antagonist views of social justice and liberal market are re-interpreted resulting in a hybrid new set of discourses with a different identity.

9.3 Shifting networks and discourses

In further elaborating on the first and second research questions, this section focuses on the shifts in networks and discourses and the changing roles of NGOs, grouped in five themes.

9.3.1 From confrontation to collaboration

Overall, this study found that over the past 20 years NGOs are becoming more collaborative and less confrontational, more conciliatory and less dogmatic, towards market actors. But, these NGOs hardly become collaborative, and remain rather conflictive and competitive, towards fellow NGOs.

The abandonment of confrontation towards the market and market actors makes NGOs not only more conciliatory and pragmatic but also more collaborative and more market-driven organizations. These changes enable the establishment of cooperative ties with SMEs, with larger companies and with other market actors. However, cooperation in the network takes place to different degrees in different networks. In the organic production networks, conflicts between conventional NGOs and market-driven NGOs and between NGOs and SMEs are more intense but their cohesion in supporting SMEs is also stronger. Agro-ecological NGOs and their affiliated organizations have strong ties and are integrated in common platforms at local, national, regional and global level. This cooperation is stimulated by commonalities in the discourses of local NGOs, SMEs and international NGOs. Regardless of their differences, all organic production networks pull together against common 'enemies', such as transgenic seed and agrochemical companies (see Chapter 5). In the business social responsibility domain, the social justice network and the business network do not share collaborative ties but they converge in their views about SMEs.

Both networks consider the role of sustainable value chain as central to improve the situation of small scale enterprises (see Chapter 6). In the third domain a different picture comes to the fore. The networks are able to cope with market and social concerns of SMEs but the differences in their discourses do not favor cross cooperation to support SMEs. Cooperation is scarce and duplications of efforts are strong among the networks of sustainable production notwithstanding the strong commonalities in most discourses (see Chapter 6).

The mismatch between network and discourse logics, with shifts in discourses towards convergence or divergence of views, is conditioning these varying degrees of cooperation. Shared views of actors are not enough to establish close ties and the presence of close ties does not guarantee collaboration. This mismatch between network and discourse logics has to do with the strong mistrust among network actors, especially between conventional NGOs and the new types of NGOs. Mistrust contributes to constructing adversarial views rather than common ground for dialogue. Such mistrust is fed by, for instance, an 'arrogant' attitude of market-driven NGOs against SMEs when the former consider most SMEs incompetent to improve their social and environmental performance. Larger companies prefer to be engaged with more 'friendly' market-driven NGOs rather than with conventional NGOs in supporting small scale providers. Even though conventional NGOs have largely abandoned their adversarial perspectives towards markets and the state, they usually are labeled as confrontational actors by the new types of NGOs.

The complementary character of discourses enhances the necessity and possibility for further integration and cooperation of network actors. For instance, solidarity and business performance, as 'communitarian' and 'individualistic' values respectively in the discourses, are both needed to improve sustainable practices in SMEs. Communitarian values enhance the mobilization and cohesion of SMEs to perform successfully in sustainable markets; and individualistic values strengthen individual entrepreneurship and competitiveness to develop and access sustainable markets. The organic production networks are the best example of the shift of NGOs towards cooperation, conciliation and flexibility. In the business social responsibility networks, the convergence of NGOs is facilitating the establishing of cooperative ties (an example of this is the collaboration of NGOs in the 'Committee Peru of the ISO 26000'). In the sustainable business networks, cross cooperation and integration of NGOs will need more time but might happen in the future, at least between those networks that share strong commonalities in their discourses such as the eco-efficiency network, the cleaner technology network, the technological innovation network and the urban cleaner production network.

9.3.2 Liberal market in the centre

The liberal market has moved to the centre in the discourses of sustainability of SMEs and it is perceived as an effective condition to push SMEs from pollution to sustainability, as long as it is merged with social and sustainability claims.

As a result of this vital and central position of the liberal market, social justice and environmental claims of NGOs have been transferred to the market arena. The market sets the boundaries for cooperation, bargaining and influence of actors. However, this transfer from the social arena to the market arena brings a tension in the discourses of NGOs between the social justice and environmental claims, as I argued in Chapter 8.

Another aspect of the liberal market is that it is acknowledged in all discourses but is not understood in the same way in all discourses. The liberal market is not only acknowledged as market 'access' but also in terms of 'adjustment' to the market and the 'democratization' of the market. This diversity of understandings of the liberal market does not mean full endorsement to the claims of the liberal market discourse or the social movement discourse but the acknowledgement of the worth of access to, adjustment to and fair relationships within the market.

Another consideration regarding the centrality of the liberal market is that it is usually seen as an enabling condition for pushing SMEs from polluting to sustainable enterprises engaged in profitable value chains; but at the same time the liberal market has its limitations in pushing the majority of SMEs that are attached to local markets towards sustainability. However, the enabling and limited conditions of the liberal market depend on the extent that it is merged with social and sustainability claims. The research findings show that global markets and decentralized local markets play a valuable role in fostering sustainable SMEs as far as they are driven by sustainability. By combining market with social and environmental views the discourses of sustainability of SMEs overcomes the dichotomy of global and local markets. Therefore, sustainable global and local markets are not conflicting but reinforcing mechanisms for improving the environmental performance of SMEs in Peru.

9.3.3 New power of SMEs

The power of SMEs is not acknowledged in most discourses. However, SMEs show their power by accepting or denying connecting to the networks, by collaborating or pressuring key actors, and by subscribing or being indifferent to the discourses. SME show their power as long as networks become more inclusive, participatory and valuable for SMEs. Power of SMEs rests on the capacity to be anchored within local social networks.

The failure to acknowledge the power of SMEs makes that SMEs are underestimated, limited in their network participation and are excluded from most networks. In the sustainable production networks, participation of SMEs is limited to being a recipient of resources. In the business social responsibility networks most SMEs, especially small and micro scale enterprises oriented to local markets, are excluded. In this domain, there is no room for SMEs in the various networks to gain power positions as they are dependent on their larger buyers. SMEs attached to exporting supply chains are the target of business social responsibility networks. A different picture emerges from the organic production networks. In this domain, small scale producers and small scale enterprises challenge the power position of conventional NGOs and international cooperation agencies. Although those actors dominate the networks, SMEs are gaining power and improve their position. This change in position of SMEs from a marginal to a more central position has to do with the rise of sustainable SMEs, the capacity of SMEs to establish their own NGOs and platforms, and the connection of SMEs to competitive national and international market actors. SMEs have even achieved direct participation in platforms at regional and global level (see Chapter 5). SMEs show their power in all the networks not only to a different degree but also in different ways. The limited motivation of SMEs to buy-in most discourses put them in a powerful position to delegitimize interventions of agents. The power of SMEs rests on their capacity to be anchored within local social networks. Acknowledging the power of SMEs implies that SMEs are not just beneficiaries but active actors with decision capacities in the networks. Without the participation of SMEs in intervention policies, the good intentions of promoters of SME sustainability will not move beyond a few 'successful' pilot experiences. NGOs need to acknowledge the views of SMEs if they really want to improve the social and environmental conditions of SMEs. Particularly, NGOs need to acknowledge the motivations of SMEs, be embedded in SMEs' local social networks and establish the conditions for SMEs to show their power in the market. By doing so, NGOs can be true nodes for bi-directional flows of resources, from global to local and local to global actors, as the case study of organic production showed.

9.3.4 New roles of NGOs

Next to the conventional 'watchdog' roles, NGOs are developing new roles of 'helper' in order to answer to the market needs of SMEs. The new roles are performed by 'reoriented' conventional NGOs and by new types of market-driven NGOs.

New roles of NGOs are market facilitation, market intermediation, capacity building of business skills and allocation of sustainable technologies for production. Market facilitation mainly includes the function of supporting the development of farmers' markets at the national level. Market intermediation mainly includes the intermediation of SMEs to access competitive markets and the intermediation between small scale suppliers and larger buyers. The function of capacity building includes providing workshops and trainings in business skills for SMEs. Finally, the function of allocating sustainable technologies to SMEs implies a greater specialization of NGOs in cleaner and appropriate technology. The new roles imply the channelling of resources in networks from international (inter)governmental organizations, international financial institutions and international NGOs to (local) SMEs. The emergence of those roles has to do with the fact that NGOs are gaining expertise in management, technology and commercialization on the one hand, and are getting tied to larger companies, SMEs or the state one the other hand. However, it is important to point out that although greater specialization of NGOs in market and technology issues is crucial, sensitivity for the social and cultural background of SMEs is also vital to engage SMEs in the networks and discourses of sustainability. The emergence of those new roles is in line with the increasing focus of SMEs on international markets. Regardless of the emergence of the aforementioned new roles, NGOs continue to use pressure and advocacy to influence larger companies and their value chains, in order to enhance the access of SMEs. This means that currently NGOs perform both 'helper' roles and 'watchdog' roles. In the new role of helper, NGOs with a more market-driven approach are becoming a new partner of companies. Conventional NGOs usually combine watchdog and helper roles in collaboration with larger companies, but market-driven NGOs usually only perform helper roles. New roles of NGOs represent not only opportunities but also risks for their legitimacy. The dilemma for conventional NGOs is to what extent they should become engaged in or keep a distance to the market and larger companies in supporting SME sustainability. Market-driven NGOs have a more clear position about their helper role but they compromise their legitimacy.

9.3.5 NGOs as market agents

Consequently, following from the former point, NGOs have become market agents as a result of their new roles.

NGOs have become main agents in promoting sustainability of SMEs by promoting eco-efficiency measures, organic certification and social and environmental standards. Although companies are much better equipped to operate effectively and efficiently on markets than NGOs, they lack capacities to connect directly with SMEs. Companies need to establish contacts or cooperate with NGOs or similar organizations to reach SMEs. It proves easier for NGOs to gain capacities to operate on markets, than for companies to gain capacities and understand the reality of SMEs. So the change of NGOs from aid agents to market agents has come as a result of the demands of market actors for building ties with SME. This means that conventional NGOs will be somewhat marginalized, and consulting or company assisting NGOs will become more central. Different from conventional aid NGOs founded in the 1980s, NGOs promoting sustainability of SMEs base their support on a market approach. Hence, NGOs have changed in structure and discourse, for instance by implementing a more standardized work style, affiliating SMEs within their organizational structure, accepting or even advocating the liberal market, and cooperating in networks of global scope. So agency capacity of NGOs, especially of local NGOs, does not primarily come from high capital resources or access to global resources and markets, but from the capacity to establish cooperation ties with other NGOs and SMEs, and to be anchored in local social networks.

9.4 Generalizability of the research findings

Before addressing the theoretical implications of this research, and thus answering the third research question, this section investigates to what extent the findings and the research methodology can be generalized to other case studies, sectors and countries.

Although the findings are built on the basis of empirical work in Peru, they have wider validity beyond Peru. This is so because the networks of sustainability of SMEs are not limited to Peru. The networks include actors from most Latin American countries and other regions where NGOs and SMEs are organized in platforms at the national level as well. For instance, to mention some countries in the region in addition to Peru: the organic production networks have branches in Bolivia, Ecuador, Brazil and Costa Rica; the business social responsibility networks have branches in Chile, Argentina and Brazil; and the sustainable production networks have branches in Chile, Colombia, Nicaragua, Cuba and Brazil. NGOs and SMEs from those countries are tied to the same regional platforms identified in the research for each domain (see Chapter 5, Chapter 6 and Chapter 7). The prominent participation of NGOs as actors in improving the conditions of SMEs is particularly strong in Latin America in comparison to other regions. The active role of Latin American NGOs in development has been highlighted by several authors such as Johnson (1998), Balbis (2001) and Wood and Roberts (2005). So, similar tendencies as those found for Peru are expected at the regional level, but not necessarily with respect to other regions such as sub-Saharan Africa, South Asia or the Middle East. Despite the fact that South Asian countries like Vietnam and China have dynamic SMEs, national NGOs do not play central roles in SME development in these countries but the state. In contrast to Africa and Asia, Latin American NGOs have more room for delivering development, are more interested in influencing politics and reach great number of beneficiaries working with and through existing local organizations (Wils, 1995). An additional factor conditioning the increasing role of NGOs in SME sustainability in Latin America is the low effectiveness of the Latin American states in delivering social and environmental improvements. In this vein, Bebbington and Thiele (1993) highlighted that NGOs in Latin America would be more than willing to fill the gaps left by a receding state. Hence, the presence of stronger civil society with well organized national and local NGOs, and more integration of national economies to global chains and networks and the failure feature of Latin American states make Latin America a region with stronger expansion of non-state actors than the aforementioned regions.

As the increasing role of NGOs in engaging with productive activities and market processes from local to global scope challenges conventional research methodologies, the combined use of social network analysis and discourse analysis can be helpful to conduct other studies about the role of NGOs. Social network analysis and discourse analysis studies have usually been conducted separately to study environmental issues. Network analysis and discourse analysis can be applied in combination to asses in an integral way NGOs promoting sustainability of SMEs worldwide in other domains, such as sustainable energy, fair trade and eco-tourism. The advantage of combined network and discourse analyses is the highly precise identification of actors and their connections, storylines and power relations that provide a deeper understanding of the role of non-state actors.

9.5 Theoretical implications

This section answers the third research question, focusing on the contribution of this study to ecological modernization theory and network society theory and to network and discourse models.

9.5.1 Network society theory

Networks promoting sustainability of SMEs help to understand network cooperation deeper but challenge Castells scheme of placing space of flows opposite to space of places.

The way actors cooperate in the networks promoting sustainability of SMEs does not contradict network society theory and general theories of networks. Furthermore, the networks investigated help to deeper understand cooperation in networks. Networks promoting sustainability of SMEs show several degrees of cooperation. These patterns make these networks more effective or less effective in reaching their proposed aims. Depending on the degree of cooperation and strength of ties in the networks, three patterns of change can be found in network configurations: a 'unidirectional shift' for the organic production networks, a 'converging shift' for the business social responsibility networks and a 'diverging shift' for the sustainable production networks. The diversity in expertise, the interdependency of actors and cohesion in their platforms contribute to making networks more stable, as the organic production networks show. Organic production networks involve a large number of NGOs, including NGOs established by SMEs, and SMEs. Oppositely, overlap of interests and lack of strong forces binding actors, causing network disintegration and isolation of actors contribute to make networks less stable as the sustainable production networks show.

Castells scheme of flows versus place does not work to understand how SME sustainability is restricted or advanced. NGOs as 'spokesmen' and interest representative of space of places and local identities take up and use also typical rationalities, logics and power resources that we conventionally find (or at least Castells relate to) space of flows. And companies, also those connected to global business networks that are part of the space of flows dynamics do and can include typical 'local' space of place rationalities, logics, dynamics and resources, such as related to sustainability. The further connection of SMEs with global economic networks is causing the taking up of new rationalities. This exchange in rationalities is currently happening in the organic production networks to a major degree and to a minor degree in the other two types of networks. As such, Castells' dichotomic scheme of flows versus place proves in reality much more complicated and intermingled when we want to understand sustainability.

Castells underestimates how global dynamics and networks can connect in a less confrontational way to local dynamics and interests, to enable, facilitate and advance interests that are usually believed to be only infringed by the space of flows. The networks of sustainability of SMEs form a good example. Local actors in the networks of sustainability of SMEs are not limited to contesting the liberal market economy. Local actors embrace the central position of liberal market economy and approach it from three viewpoints: adaptation, access and fair relationship. So, local actors adapt to, access and democratize the neoliberal market economy.

Local and global actors have not only opposing views organized in power and counter power as Castells claims, but also shared views as they endorse common discourses (see Chapter 5, Chapter 6 and Chapter 7). Rather than local actors just becoming nodes of 'alternatives to global networks', they are embedded and integrated in global conventional networks, to the extent that it is no longer useful to distinguish two opposing spheres, of flows and place, but seeing them connected and integrated in aiming for sustainability. Therefore, networks promoting sustainability of SMEs can be seen as hybrid networks involving actors (e.g. larger companies, NGOs, SMEs and government agencies) from local to global level as they crosscut 'conventional' market networks, civil society networks and governmental networks. This hybrid characteristic of the networks promoting sustainability of SMEs is facilitated by close interaction of NGOs and SMEs.

Finally, local actors are not powerless in the dominant networks, as Castells claim. While global actors intend to control local actors, local actors try to influence global actors and emancipate from them. Thus, SMEs hold power, as international NGOs do. While international NGOs and other global actors hold 'informational' power, SMEs and other local actors hold 'anchoring' power (see Chapter 8). In the networks any actor can exercise power to influence or emancipate from other actors. SMEs in the networks have the capability to legitimate or challenge the power of international cooperation agencies and local NGOs. This 'anchoring' power is a key resource of SMEs to gain a position in networks, and bargain and negotiate with international cooperation agencies and NGOs by assembling, influencing, refusing or endorsing discourses. Anchoring power constitutes an advantage of SMEs not only to gain a position in the networks under conditions of globalization but to become a source of 'fresh' views beyond the dichotomy of liberal market and social movement. The power of SMEs is realized in three developments. First, SMEs (e.g. BioFrut and La Cabrita) are gaining power in the networks as they are getting connected to profitable markets but keeping their roots in localities. Second, SMEs build up their own organizations (e.g. Bioferias) under conditions in the networks the enable participation of SMEs. And third, the emergence of new NGOs representing small scale enterprises (e.g. ANPE, CEPICAFE, APROMALPI, CEPIBO and REPEBAM) provides SMEs additional influence and power.

9.5.2 Ecological modernization theory

Non-state actors perform central roles in a more 'inclusive' environmental governance of SMEs, quite in line with the central propositions of ecological modernization theory. However, ecological modernization has to take into account the mutual dependence of ecological rationality and social rationality in advancing environmental reform of SMEs in developing countries.

The empirical discourses of sustainability of SMEs and the theoretical discourse of ecological modernization have similarities and differences on how to deal with polluting SMEs. Both discourses share views on the secondary role of the state to eco-modernize industries. In the discourses of sustainability of SMEs, NGOs and SMEs constitute the main actors in planning, channeling resources and implementing projects in SMEs to improve their environmental performance *vis-à-vis* their economic concerns. The state is virtually absent in the networks studied or plays only a secondary role. If more central roles are played by state authorities usually they fail, for instance in the sustainable production networks (see Chapter 7). Both discourses share views on a deficient role of the traditional bureaucratic state in environmental reform.

One of the main differences is the fact that in this study NGOs take a strong position than market agents in promoting sustainability of SMEs. NGOs are central actors in the networks of sustainability of SMEs, not companies. The theory of ecological modernization usually articulates this preferential position and role to companies, and not so much to NGOs. As market agents NGOs perform several roles not only to support SMEs to access global sustainable markets but also to develop local sustainable supply chains. The emerging of new types of NGOs in networks such as producer NGOs, market NGOs, business NGOs and technocratic NGOs, has been an answer to the market demands of SMEs. This wide-ranging role of NGOs put them in the front-line promoting environmental sustainability of SMEs in comparison to larger companies and the state.

A second relevant difference is that the discourses of sustainability of SMEs include social and employment conditions and rationalities in defining and advancing sustainability of SMEs, whereas ecological modernization is restricted to articulating ecological rationality. NGOs and SMEs in our empirical networks go beyond ecological rationality. The discourses of sustainability of SMEs call not only for the introduction of sophisticated and tailored environmental methods but also for the application of social and environmental interlinked interests, such as social responsibility standards, organic certification and participatory guaranty systems. Dealing with polluting SMEs and the concerns of sustainability and market access imply the use of the environmental rationality linked to social and employment conditions and rationalities. This difference is especially evident in the market justice discourses (see Chapter 8). In this way, the discourses of sustainability of SMEs enrich the perspective of ecological modernization in approaching environmental reform based on a single rationality – environmental rationality – towards an interlinked application of social and environmental rationalities, especially to advance environmental reform of SMEs in developing countries.

9.6 Recommendations

This section addresses a set of suggestions for governance and research in the sustainability of SMEs worldwide.

9.6.1 Recommendations for governance

Three governance recommendations follow from this research. This research suggests that increased cooperation among NGOs would contribute to more effective support of SMEs. There is a need to cross-cooperation of forefront NGOs and SMEs affiliated to organic production, business social responsibility, sustainable production and other networks promoting sustainable SMEs (e.g. ecotourism, sustainable energy and fair trade). One of the ways to achieve such cooperation at national and sub-national level is the establishment of coordination boards, possibly based on the platforms already established. It is of particular importance to have boards which are not confined to one sector but reach over different sectors. One of the main tasks on the coordination board agenda has to be encouraging SME organizations (e.g. associations, cooperatives and platforms) to commit towards sustainability.

Market actors can play significant roles in promoting sustainability among SMEs. It is recommended that sustainable larger companies help to pull polluting small scale supplier companies towards more sustainable practices, for instance by providing technical and marketing support. SMEs in the supply chain should be more aware of the power they can derived from their local embedding and creativity, a power that can also be used to move larger companies towards more sustainable value chains. Consumers and consumer organizations, quite obviously, can play an important role in promoting sustainable SMEs by the power of consumer demand.

The networks of sustainability of SMEs are a source of learning for policy makers to overcome the actual failures in national environmental policy in tackling polluting SMEs and promoting sustainable SMEs. This research shows that NGOs and their platforms in the networks are more effective in encouraging SMEs towards sustainability that the national government. Therefore, national governments should elaborate national policies for environmental governance of SMEs based on the experiences of the networks of NGOs rather than elaborate new regulations and build up parallel structures of SMEs governance. This recommendation is particularly applicable to Latin American national governments as the networks operate along the region.

9.6.2 Recommendations for research

Three research recommendations can be mentioned. One essential issue on the research agenda of promoting sustainability of SMEs throughout Latin America would be the deeper understanding of cooperation and power relations between NGOs and SMEs. Carrying out comparative cross-country studies, pointing out similarities and differences regarding the shifting patterns and tendencies identified in the Peruvian cases, would be interesting for that. Central aim of these studies would be to find out participation and decision capacity of SMEs in the sustainability network branches in other countries such as Bolivia, Brazil, Chile and Costa Rica. Also further research whether the discourses promoting sustainability of SMEs identified in the Peru are endorsed by NGOs and SMEs in these other countries is interesting.

A second issue on the research agenda would be to generate knowledge on the cooperation mechanisms between SMEs, large companies and market facilitating NGOs involved in sustainable supply chains and sustainable markets. Central in these studies would be to identify the enabling and the constraining factors for sustainable supply chains. As sustainable global and sustainable local markets are reinforcing strategies for sustainability, these studies should focus on the social, political and economic dynamics of global value chains. The increasing dependencies of social, economic and cultural issues in SMEs and larger companies influence the improvement of environmental performance of business. Therefore, we recommend to not focus on SMEs alone, but to include both local linkages with other social, cultural, and economic processes as well as global value chains as important context factors.

Based on the methodological contribution of this research, studies are recommended that show to what extent the network-discourse model is applicable for other domains such as sustainable energy, fair trade and eco-tourism, highlighting the role of NGOs and SMEs and the discourses involved in these domains.

References

Aburto, J. (2006). Entrevista a Henri Le Bienvenu, gerente general de Perú 2021. Año 1, No 0. Pag 18. *CANALE Comunicación + desarrollo.* Available at: http://revistas.pucp.edu.pe/index.php/canale/article/view/1454/1400.

Alegre, A., & Marthaler, C. (2010). *Territorios eco-eficiencies – ecoparques industriales del Callo.* CER. GEA. Lima, Perú. GyG Impresores SAC.

Alegre, M. (2007). *Ecoeficiencia y rentabilidad empresarial. Casos prácticos nacionales e internacionales.* Centro de eco-eficiencia y responsabilidad social (CER). SECO: Lima, Perú.

Alegre, M. (2009). Ecoeficiencia en las empresas y el cambio climático. Experiencias prácticas. CER. GEA. Available at: http://www.camaralima.org.pe/bismarck/DESCARGAS/ppt%20MINAM%20CCL_Marcos%20Alegre.ppt.

Alvarado, F. (2002). La agricultura ecológica: conservación de la biodiversidad y mercado. Centro IDEAS, RAE-PERÚ.

Andersen, M.S., & Massa, I. (2000). Ecological modernization – origins, dilemmas and future directions. *Journal of Environmental Policy and Planning,* 2, 337-345.

Anheier, H. K., & Katz, H. (2003). Mapping global civil society, pp. 241-258. In: M. Kaldor, H. Arheier and M. Glasius (eds.), *Global Civil Society* 2003. Oxford University Press: Oxford, UK.

Anheier, H. K., & Katz, H. (2004). *Network approaches to global civil society.* Global Civil Society 2004/5. The London School of Economics and Political Science (LSE), University of California. SAGE Publications: Thousands Oaks, CA, USA.

Arbulú, J. (2007). PYME: La socia mayoritaria del Perú. *Diario El Peruano* April 13, 2007.

Arts, B. (2002). Green alliances of business and NGOs. New styles of self-regulation or dead-end roads? *Corporate Social Responsibility and Environmental Management Journal,* 9, 26-36.

Balbis, J. (2001). NGOs, governance and development in Latin America and the Caribbean. Management of Social Transformations – MOST. Discussion Paper No. 53.

Banco Mundial (2007). *Análisis ambiental del Perú: retos para un desarrollo sostenible.* Resumen Ejecutivo. Unidad de Desarrollo Sostenible Región de América Latina y el Caribe. Banco Mundial. Perú. Tarea Asociación Grafica Educativa.

Bebbington, A., & Thiele, G. (1993). *Non-governmental organizations and the state in Latin America: rethinking roles in sustainable agricultural development.* Routledge: London, UK.

Bebbington, A., Rojas. R., & Hinojosa, L. (2002). Contributions of the Dutch co-financing program to rural development and rural livelihoods in the highlands of Peru and Bolivia. Synthesis report. Study commissioned by: Steering Committee for the Evaluation of the Netherlands' Co-financing Programme.

Beloe, S., Elkington, J., Fry Hester, K., Newell, S., Kell, G., & Aloisi, J. (2003). *The 21st century NGO: in the market for change.* SustainAbility/United Nations Environment Programme: London, UK.

Benton, L.M., & Short, J. R. (1999). *Environmental discourse and practice.* Blackwell publishers: Oxford, UK.

Benzing, A. (2001). *Agricultura organica. Fundamentos para la region andina.* COSUDE, ERPE, MISEREOR, GTZ/ IICA, SWISSAID, Universidad de Kassel. Neckar-Verlag: Villingen-Schwenningen, Germany.

Bianchi, R., & Noci, G. (1998). Greening SMEs' competitiveness. *Small Business Economics Journal,* 11, 269-281.

Biekart, K. (2005). *Políticas de las ONG europeas para América Latina: tendencias y perspectivas recientes.* ICCO – ALOP: The Hague, the Netherlands.

Blowfield, M., & Frynas, J.G. (2005). Setting new agendas: critical perspectives on corporate social responsibility in the developing world. *Journal of International Affairs,* 81(3), 499-513.

Brand, E.M.L., & De Bruijn, T.J.N.M. (1999). Shared responsibility at the regional level; the building of sustainable industrial estates. *Journal of European Environment*, 9(6), 221-231.

Bridge, S., O'Neill, K., & Martin, F. (2009). *Understanding enterprise: entrepreneurship and small business*. Third edn. Palgrave Macmillan: Basingstoke, UK.

Brilhante, O. (2001). Environmental management systems (EMS) and pollutions prevention in small and medium enterprises (SMEs) in the Netherlands, Brasil and Vietnam. HIS: Rotterdam, the Netherlands.

Brio, J.A., & Junquera, B. (2003). A review of the literature on environmental innovation management in SMEs: implications for public policies. *Technovation Journal*, 23, 939-948.

Burga, R. (2009). Impacto climático y ecoeficiencia. *Revista Stakeholders*, 3(16), 22-23.

Buttel, F. H. (2003). Environmental sociology and the explanation of environmental reform. *Organization Environment Journal*, 16(3), 306-344.

Buttel, F.H. (2001). Environmental sociology and the explanation of environmental reform. Paper presented at *the Kyoto Environmental Sociology Conference*, Kyoto, Japan, 21-23 October 2001.

Canessa, G., & Garcia, E. (2005). *El ABC de la responsabilidad social empresarial en el Peru y el mundo*. Peru 2021. Siklos. S.R. Ltda: Lima, Peru.

Caravedo, B. (2008). Responsabilidad social: TODOS. *Revista Stakeholders,* 2(11), 28-33.

Caravedo, B. (2011). Entrevista. La responsabilidad social es una energía cohesionadora. *Revista Stakeholders*, 28, 28-29.

Castells, M. (1997). *The power of identity, the information age: economy, society and culture*. Vol. II. Blackwell: Oxford, UK.

Castells, M. (2000). *The rise of the network society*. 2nd. edition. Blackwell publishing: Oxford, UK.

Castells, M. (Editor) (2004). *The Network society. A cross-cultural perspective*. Edward Elgar Publishing: Cheltenham, UK.

Cavero, E. (2007). Orgánicos: la invasión verde. *Diario El Comercio* March 18, 2007.

Centro IDEAS (2011). Boletín de IDEAS No 4. Enero 2011. Centro IDEAS. Lima, Perú. Available at: http://servindi. org/pdf/Boletin_de_IDEAS_041.pdf.

Cerin, P. (2004). Where is corporate social responsibility actually heading? *Progress in Industrial Ecology*, 1, Nos. 1/2/3.

Chavalparit, O. (2006). *Clean technology for the crude palm oil industry in Thailand*. PhD Thesis Wageningen University, Wageningen, the Netherlands.

Christie, I., Rolfe, H., & Legard, R. (1995). *Cleaner production in industry*. PSI Publishing: London, UK.

Cici, C., & Ranghieri, F. (2008). *Recommended actions to foster the adoption of corporate social responsibility (CSR) practices in small and medium enterprises (SMEs)*. IDB-RGA: Washington, DC, USA.

Cohn, A., Cook, J., Fernández, M., Reider, R., & Steward, C. (eds.). (2006). *Agroecology and the struggle for food sovereignty in the Americas*. Institute for environment and development (IIED), the IUCN commission on environment, economic and social policy (CEEPS) and the Yale school of forestry & environmental studies (Yale F&ES).

Comisión Económica para América Latina y el Caribe (CEPAL) (2006). Insumos para identificar políticas innovadoras, lecciones aprendidas y mejores prácticas en los temas de energía, desarrollo industrial, contaminación del aire/ atmosfera y cambio climático en la región de América Latina y el Caribe. *Talle regional sobre la Decimo Quinta Sesión de la Comisión sobre el Desarrollo Sostenible de las Naciones Unidas (CDS 15)*. Ciudad de Mexico, Mexico, 7 y 8 de Setiembre de 2006. División de Desarrollo Sostenible y Asentamientos Humanos de la CEPAL, Naciones Unidas.

Comisión Económica para América Latina y el Caribe (CEPAL) (2010). *Políticas de apoyo a las pymes en América Latina. Entre avances innovadores y desafíos institucionales*. Compiladores: Carlo Ferraro and Giovanni Stumpo. CEPAL. Col. Libros de la CEPAL, No 107, Cooperazione Italiana. Santiago, Chile.

Compartiendo (2012). Boletin electrónico semanal. Marzo. Elaborado por Fernando Alvarado De La Fuente. Centro IDEAS. No 10/ No 11.

Condor, R. (2001). *Towards a sustainable microenterprise sevelopment: From an industrial park to an eco-industrial park, the case of Villa El Salvador, Lima, Peru.* Master thesis. Wageningen University, Wageningen, the Netherlands.

Consejo Nacional de Ciencia y Tecnología del Perú (CONCYTEC) (2006). *Curso taller transferencia de tecnologías limpias para PYMEs del sector residuos sólidos.* 23 al 25 de octubre de 2006. OPS. OEA. Lima, Perú.

Consejo Nacional de Ciencia y Tecnología del Perú (CONCYTEC) (2009). Plan nacional de ciencia, tecnología e innovación tecnológica para el desarrollo productivo y social sostenible 2009-2013. SINACYT. Lima, Perú.

Corral, A., Isusi, I., Peinado-Vara, E., & Pérez, T. (2005). *La responsabilidad social y medioambiental de la microempresa en Latinoamérica.* IKEI. IDB.

Cox, J. (2006). ¿Cual es tu propósito? DESAFIOS – Hacia una nueva cultura empresarial. No 68. Abril-Mayo 2006. CSIRO Publishing: Collingwood, Australia.

Cuneo, F. (2005). La RSE es una filosofía social y no una moda. *Punto de Equilibrio,* 14 (88), Julio.

Cuneo, F. (2007). La empresa como agente de cambio. Ponencia. Diapositivas en power point. Perú 2021.

De la Torre, D. (2004). Más allá de lo que se ve. Especial. Responsabilidad Social Empresarial. Negocios Internacionales.

Diani, M., & McAdam, D. (eds.) (2003). *Social movements and networks: relational approaches to collective action.* Oxford University Press: Oxford, UK.

Dias, D., Filomeno, M., & Rizo-Patron, C. (2007). *Relación y compromiso con los grupos de interés.* Guía práctica para las empresas peruanas. Hacia la responsabilidad social estratégica. Apoyo Comunicación Corporativa. Perú 2021.

Diaz-Albertini, J. (2001). La participación política de las clases medias y las ONGDs en el Perú de los noventa. *América Latina Hoy,* Vol. 28. Instituto de Estudios de Iberoamérica y Portugal, Universidad de Salamanca, Salamanca, Spain.

Dryzek, J. (1997). The Politics of the earth: environmental discourses. Oxford University Press: Oxford, UK.

Dutch interchurch organization for development cooperation (ICCO) (2009). ICCO position paper on business social responsibility. Available at: http://iccointwebsite.prd.prisma-it.com/int/linkservid/9D20549B-F759-0A0D-0DC3118F1ACF1ECC/showMeta/0/.

Dutch interchurch organization for development cooperation (ICCO) (2010). ICCO alliance annual report 2010. Available at: http://m.icco.nl/documents/pdf/icco_alliance_annual_report_2010_-_lowres.pdf.

Dutch interchurch organization for development cooperation (ICCO) (2011). ICCO annual report 2011. Available at: http://www.icco.nl/nl/linkservid/51BB894E-DA2D-7604-F26CAE09019477C5/showMeta/0/.

Ebrahim, A. (2005). *NGOs and organizational change. Discourse, reporting, and learning.* Cambridge University Press.

Espinoza, W. (1987). *Los incas. Economía, sociedad y estado en la era del Tahuantinsuyo.* Amaru Editores: Lima, Perú.

Febres, M. (2004). Informe regional desk – sector industrial. Proyecto Regional Desk de la Red de Transferencia Tecnológica. CONCYTEC.

Felipe-Morales, C. (2000). Agroecológica. Un enfoque para nuevas políticas. La Revista Agraria No 20. Lima, Perú, octubre 2000. Available at: http://www.larevistaagraria.info/sites/default/files/revista/r-agra20/arti-01b.htm.

Ferraro, C., & Stumpo, G. (2010). Políticas de apoyo a las pymes en América Latina: Entre avances innovadores y desafíos institucionales. CEPAL.

Flores, G. (2004). Lineamientos para el establecimiento del nodo regional andino de TTN. Informe preparado para la primera reunión Andina. Documento TTN/Andino/001 Rev.1. CONCYTEC-PNUMA.

Food and Agriculture Organization of the United Nations (FAO) – Regional Office for Latin America and the Caribbean (RLC) (2009). *Panorama de la seguridad alimentaría y nutricional en América Latina y el Caribe.* FAO, Observatorio del Hambre, Iniciativa América Latina sin hambre.

Forum Empresa (2009). Reporte de gestión 2009. Secretaría general de la red. Forum Empresa.

Foundation for Research on Multinational Corporations (SOMO) (2009). SOMO annual report 2009. Available at: http://somo.nl/publications-en/Publication_3602.

Frijns, J., & van Vliet, B. (1999). Small-scale industry and cleaner production strategies. *World Development Journal*, 27(6), 967-983.

Frijns, J., Kirai, P., Malombe, J., & Van Vliet, B. (1997). *Pollution control of smallscale metal industries of Nairobi*. Department of Environmental Sociology, Wageningen Agricultural University: Wageningen, the Netherlands.

Garcia, A. (2007). El sindrome del perro del hortelano. *Diario El Comercio*. Available at: http://elcomercio.pe/edicionimpresa/html/2007-10-28/el_sindrome_del_perro_del_hort.html.

Gärdström, T., & Norrthon, P. (1994). Implementation of cleaner production in small and medium-sized enterprises. *Journal of Cleaner Production*, 2(3-4), 201-205.

Gombault, M., & Versteege, S. (1999). Cleaner production in SMEs through a partnership with (local) authorities: successes from the Netherlands. *Journal of Cleaner Production*, 7, 249-261.

Grabher, G. (1993). Rediscovering the social in the economics of interfirm relations. Routledge, London. In: Grabher, G. (ed.), *The embedded firm – On the socioeconomics of industrial networks*. Routledge: London, UK, pp. 1-31.

Granovetter, M. (1983). The strength of weak ties: a network theory revised. *Sociological Theory Journal*, 1, 201-233.

Grohmann, P. (1997). Los movimientos sociales y el medio ambiente urbano. *Nueva Sociedad* 149 (Mayo-Junio), 146-161.

Hajer, M., 1995. *The politics of environmental discourse: ecological modernisation and the policy process*. Oxford University Press: Oxford, UK.

Halberg, N., Alrøe, H.F., Knudsen, M.T., & Kristensen, E.S. (eds.). (2006). *Global development of organic agriculture. Challenges and prospects*. Danish Research centre for organic and food farming. CABI Publishing: Wallingford, UK.

Hall, C. (2010). *United nations global compact annual review 2010*. The United Nations Global Compact Office: New York, NY, USA.

Hallstrom, L. (2004). Eurocratising enlargement? EU elites and NGO participation in European environmental policy. *Environmental Politics Journal*, 13(1), 175-193.

Hechaverria, A. (2004). Informe regional desk – sector textil y confecciones. Proyecto Regional Desk de la Red de Transferencia Tecnológica. CONCYTEC.

Holt, G., & Reed, M. (eds.). (2006). *Sociological perspective of organic agriculture: from pioneer to policy*. CAB International: Wallingford, UK.

Hotta, Y. (2004). *The transnational politics of ecological modernization, an analysis of the formation of transnational authority in global environmental and industrial governance, with special reference to the zero Emissions Initiative in Japan*. PhD in International Relations thesis. the University of Sussex: Sussex, UK.

Huber, J. (2000). Towards industrial ecology: sustainable development as a concept of ecological modernization. *Journal of Environmental Policy and Planning*, 2, 269-285.

Humanist Institute for Cooperation (HIVOS) (2011). HIVOS annual report 2011. Available at: http://www.hivos.org/sites/default/files/annualreport2011_0.pdf.

Hyman, E., & Dearden, K. (1998). Comprehensive impact assessment systems for NGO microenterprise development programs. *World Development Journal*, 26(2), 261-276.

Inter-American Development Bank (IDB) (2008). *IDB country strategy with Peru 2007-2011*. Inter-American Development Bank: Washington, Dc, USA.

Intermediate Technology Development Group (ITDG) (2009). Memoria institucional 2007-2009. Tecnologías desafiando la pobreza. Soluciones Practicas ITDG: Lima, Perú.

Jaffe, A., Newell, R., & Stavins, R. (2003). Technological change and the environment. Chapter 11. *Handbook of environmental economics*. Vol. 1. Elsever Science BV Available at: http://www.hks.harvard.edu/fs/rstavins/Papers/Technological_Change_and_the_Environment_Handbook_Chapter.pdf.

Jennifer J. (1998). New social actors in environment and development in Latin America. In: *Cultural Environments & Development Debates, Latin America*. Regional worlds Programme. University of Chicago: Chicago, IL, USA, pp 69-74.

Jiménez, F. (2001). El modelo neoliberal peruano: límites, consecuencias sociales y perspectivas. En: *El ajuste estructural en América Latina*. Capítulo 7. CLACSO. Available at: http://bibliotecavirtual.clacso.org.ar/ar/libros/sader/cap07.pdf.

Kaimowitz, D. (1993). The role of nongovernmental organizations in agricultural research and technology transfer in Latin America. *World Development Journal*, 21(7), 113-1150.

Korhonen, J. (2000). Completing the industrial ecology cascade chain in the case of a paper industry – SME potential in industrial ecology. *Eco-Management and Auditing Journal*, 7(1), 11-20.

Korten, D. (1987). Third generation NGO strategies: a key to people-centered development. *World Development Journal*, 15, 145-159.

Kristiansen, P., Taji, A., & Reganold, J. (2006). *Organic agriculture. A global perspective*. omstock Publishing Associates: Ithaca, NY, USA.

Landtm, L. (1987). Non-governmental organizations in Latin America. Instituto de Estudos da Religião, Rio de Janeiro, Brazil. *World Development Journal*, 15, 29-38.

Langhelle, O. (2000). Why ecological modernization and sustainable development should not be conflated. *Journal of Environmental Policy and Planning*, 2, 303-322.

Larson, A. (2000). Sustainable innovation through an entrepreneurship lens. *Business Strategy and the Environment Journal*, 9(5), 304-317.

Le Bienvenu, H. (2007). La responsabilidad social como herramienta de competitividad. Desarrollo Sostenible. Especial. Negocios Internacionales.

Lee, Y. (1998). Intermediary institutions, community organizations, and urban environmental management: the case of three Bangkok slums. *Word Development Journal*, 26(6), 993-1011.

Liebenthal, A., & Salvemini, D. (2011). Promoting environmental sustainability in Peru: a review of the world bank group's Experience (2003-2009). IEG Working paper 2011/1. IFC. MIGA. World Bank: Washington, DC, USA.

Litfin, K. (1994). *Ozone Discourses*. Columbia University Press: New York, NY, USA.

Lusiani, N., & Feeney, T. (2009). Advocacy guidance on business and human rights in the United States. Part I: the mandate of the special representative. International Network for Economic, Social and Cultural Rights (ESCR-Net). Corporate Accountability Working Group.

Mead, D., & Liedholm, C. (1998). The dynamics of micro and small enterprises in developing countries. *World Development Journal*, 26(1), 6 I-74.

Minaya, E. (2007). *La responsabilidad social en el ámbito del trabajo en las PYMEs*. Red Puentes: Lima, Perú.

Ministerio de Ambiente (MINAM) (2010). Agenda nacional de investigación científica en cambio climático 2010-2021. MINAM. SINIA. Available at: http://sinia.minam.gob.pe//index.php?accion=verElemento&idElementoInformacion=287&idformula=&idTipoElemento=2&idTipoFuente=&verPor=tema.

Miranda, J. (2005). *Impacto económico en la salud por contaminación del aire en Lima metropolitana*. CIES. IEP.

Mol, A.P.J. (1995). *The refinement of production. Ecological modernisation theory and the chemical industry*. Jan van Arkel/International Books: Utrecht, the Netherlands.

Mol, A.P.J. (2000). The environmental movement in an era of ecological modernisation. *Geoforum* 31, 45-56.

Mol, A.P.J. (2001). *Globalization and environmental reform: The ecological modernization of the global economy*. MIT Press, Cambridge, MA, USA.

Mol, A.P.J., & Spaargaren, G. (2000). Ecological modernization theory in debate: a review. *Environmental Politics Journal*, 9(1), 17-50.

Mühlhäusler, P., & Peace, A. (2006). Environmental discourse. *Annual Review of Anthropology Journal*, 35, 457-479.

Murra, J. (2000). *El mundo andino. Población, medio ambiente y economía*. IEP/PUCP. Fondo Editorial Pontificia Universidad Católica del Perú e Instituto de Estudios: Lima, Perú.

Noble, R. (2000). *Global civil society and the environmental discourse: the influence of global NGOs and environmental discourse perspectives in the UNCED's declaration of principles and agenda 21*. Master thesis. International University of Japan: Minami Uonuma-shi, Japan.

Organisation for Economic Co-operation and Development (OECD) (1995). *Promoting cleaner production in developing countries*. The role of development co-operation. OECD documents.

Oxfam-Novib (2011a). Oxfam-Novib Annual Report 2011. Available at: http://www.oxfamnovib.nl/Redactie/Downloads/Rapporten/0117%201020%20Annual%20Review%202011_LOW.pdf.

Oxfam-Novib (2011b). Oxfam-Novib corporate plan 2011-2015. Available at: http://www.oxfamnovib.nl/Redactie/Downloads/Jaarverslagen/OxfamNovibsCorporatePlan%202011-2015-jun11.pdf.

Paradigmas (2010). CONCYTEC: 2006-2010 – construyendo la gestión del conocimiento, Vol. 9, No 13. CONCYTEC.

Pearce, J.M. (2012). The case for open source appropriate technology. *Environment, Developmenta nd Sustainability Journal*, 14, 425-431.

Peinado-Vara, E., & De la Garza, G. (2007). V Conferencia Interamericana sobre responsabilidad social de la empresa (RSE) responsabilidad compartida. Anales. Guatemala 9-11 de diciembre 2007.

Peres, W., & Stumpo, G. (2000). Small and medium-sized manufacturing enterprises in Latin America and the Caribbean under the new economic model. *World Development Journal*, 28(9), 1643-1655.

Peres, W., & Stumpo, G. (2002). *Pequeñas y medianas empresas industriales en América Latina y el Caribe*. CEPAL/Siglo XXI: México D.F., México.

Peru Innova (2008). Boletín No 27 – Junio 2008. Ministerio de la Producción. Red CITIEs. Available at: http://www.cites.pe/uploads/Boletines/BOLETIN%20JUNIO%20CITEs%20200820090819.pdf.

Peru Innova (2010). Boletin electrónico No 56 – Noviembre 2010. Ministerio de la Producción. Red CITIEs. Available at: http://www.cites.pe/uploads/Boletines/boletin_cites_nov0820090819.pdf.

Phung Thuy Phuong (2002). *Ecological modernisation of industrial estates in Vietnam*. Wageningen University: Wageningen, the Netherlands.

Pimenova, P., & Van der Vorst, R. (2004). The role of support programmes and policies in improving SMEs environmental performance in developed and transition economies. *Journal of Cleaner Production*, 12, 549-559.

Pradhan, R. (1989). On helping small enterprises in developing countries. *World Development Journal*, 17(1), 157-159.

Prieto-Carrón, M, Lund-Thomsen, P., Chan, A., & Muro, A. (2006). Critical perspectives on CSR and development: what we know, what we don't and what we need to know. *Journal of International Affairs*, 82(5), 977-987.

Programa de Apoyo a la Micro y Pequeña Empresa en el Perú (APOMIPE) (2007). *Integración de RSE en el trabajo de APOMIPE*. MINKA. Inter-cooperation. COSUDE. Ministerio de Trabajo: Lima, Perú.

Programa de Naciones Unidas para el Medio Ambiente – Consejo Nacional de Ciencia y Tecnología del Perú (PNUMA-CONCYTEC) (2004). Informe de avance. Proyecto de redes de transferencia de tecnología (TTN). Programa de Naciones Unidas para el Medio Ambiente – División de Tecnología, Industria y Economía (PNUMA-DTIE). Consejo Nacional de Ciencia y Tecnología del Perú (CONCYTEC). Nodo Regional: Perú.

Red de Acción en Agricultura Alternativa (RAAA) (2007). Diagnostico sobre la situación de la agricultura orgánica/ ecológica en el Perú. Unidad de Incidencia Política: Lima, Perú.

Rossi, M., Brown, H., & Baas, L. (2000). Editorial leaders in sustainable development: how agents of change define the agenda. *Business Strategy and the Environment*, 9, 273-286.

Rostworowski, M. (1999). *Historia del Tahuantinsuyo*. 2ª. Ed. IEP/Prom Perú. IEP Ediciones: Lima, Perú.

Sanborn, C., & Portocarrero, F. (eds.) (2008). *Filantropía y cambio social en América Latina*. Centro de Investigación de la Universidad del Pacifico (CIUP) and Harvard University: Lima, Perú.

Sanchez, J. (2008). *Manual de referencias sobre tecnologías apropiadas*. Convenio Andres Bello (CAB). Instituto de Transferencia de tecnologías Apropiadas para Sectores Marginales (ITACAB). Lima, Perú.

Schumacher, E.F. (1983). *Lo Pequeño es hermoso*. Titulo original: small is beautiful. Ediciones Orbis: Barcelona, Spain.

Schwalb, M., & Garcia, E. (2004). Instrumentos y normas para evaluar y medir la responsabilidad social empresarial. Documento de trabajo 66. Centro de Investigación de la Universidad del Pacifico (CIUP). Lima, Perú.

Schwalb, M., & Malca, O. (2004). *Responsabilidad social: fundamentos para la competitividad empresarial y el desarrollo sostenible*. Centro de Investigación de la Universidad del Pacifico (CIUP), Los Andes – Yanacocha: Lima, Perú.

Seippel, Ø. (2000). Ecological modernization as a theoretical device: strengths and weaknesses. *Journal of Environmental Policy and Planning*, 2, 287-302.

Sibille, A. (2004). Proyecto TTN red de transferencia tecnológica. Informe de Consultoría Sector Forestal. Programa de las Naciones para el Medio Ambiente (PNUMA). Consejo Nacional de Ciencia y Tecnología (CONCYTEC).

Soluciones Prácticas (2011). Memoria institucional 2010-2011. Usamos la tecnología para cambiar el mundo. Soluciones Prácticas. Lima, Perú. Oleograf. De Freddy Cruz.

Soluciones Prácticas (2012). Memoria institucional 2011-2012. Usamos la tecnología para cambiar el mundo. Soluciones Prácticas. Lima, Perú. Forma e Imagen.

Spaargaren, G. (2000). Ecological modernization theory and the changing discourse on environment and modernity. In: Spaargaren, G., Mol, A.P.J. and Buttel, F.H. (eds.) *Environment and Global Modernity*. Sage: London, UK, pp. 41-73.

Spaargaren, G., & Mol, A.P.J. (1992). Sociology, environment and modernity: ecological modernization as a theory of social change. *Society and Natural Resources Journal*, 5(4), 323-344.

Spaargaren, G., Mol A.P.J., & Bruyninckx, H. (2006). Introduction: governing environmental flows in global modernity. In: G. Spaargaren, A.P.J. Mol and F.H. Buttel (eds.), *Governing Environmental Flows. Global Challenges for Social Theory*, MIT: Cambridge, MA, USA, pp. 1-36.

Stafford E, Polonsky, M., & Hartman C. (2000). Environmental NGO – business collaboration and strategic bridging: a case analysis of the Greenpeace-Foron alliance. *Business Strategy and the Environment Journal*, 9, 122-135.

Su, J. (2004). Estudio para identificar sectores nacionales prioritarios. El caso de Perú. Programa de Cooperación Horizontal en Tecnologías Limpias y Energía Renovables. Proyecto OEA No SEDI/AICD/ME/221/03.

Sunkel, O. (2005). El desarrollo de América Latina ayer y hoy. Cuadernos del Cendes. Entrevista. CDC Vol. 22, No 60. Caracas, Venezuela.

Technological Transfer Network (TTN)-Red Andina (2005). Acta primera reunión andina para la transferencia de tecnologías limpias. Lima, Perú.

Teszler, R. (1993). Small-scale industry's contribution to economic development. In: I.S.A. Baud and G.A. De Bruije (eds.), *Gender, small-scale industry and development policy*. IT Publications: London, ULK, pp. 16-34.

Thorpe, B. (1994). The role of NGOs and the public to promote cleaner production. *Journal of Cleaner Production*, 2, 153-162.

Torres, V. (2004). La agricultura peruana en los tiempos del TLC. *Documento de Trabajo.* Confederación Campesina del Perú.

Tran Thi My Dieu (2003). *Greening food processing industry in Vietnam.* Dissertation Wageningen University: Wageningen, the Netherlands.

Transparency International (2005). Report on the transparency international global corruption. Barometer, (Berlin: Policy and Research Department Transparency International –International Secretariat).

United Nations Environment Programme (UNEP) (1996). *Taking action. An environmental guide for you and your community.* UNEP.

United Nations Environment Programme (UNEP) (2003). GEO Latin America and the Caribbean: environment outlook 2003. UNEP Regional Office for Latin America and the Caribbean. Mexico, D.F., Mexico.

United Nations Environment Programme (UNEP) (2010). Terminal evaluation of the UNEP / GEF project technology transfer networks Phase II: prototype verification and expansion at the country/regional level project No. GEF/2328, pp. 2740-4343.

United Nations Industrial Development Organization – United Nations Environment Programme (UNIDO-UNEP) (2010). Taking stock and moving forward. The UNIDO–UNEP National Cleaner Production Centres.

Uribe-Echevarria, F. (1991). Small-scale manufacturing and regional industrialization: the urban and regional development perspective. ISS Working Paper Series / General Series. International Institute of Social Studies of Erasmus University (ISS), pp. 1-54.

Vakil, A. (1997). Confronting the classification problem: toward a taxonomy of NGOs. *World Development Journal,* 25(12), 2057-2070.

Van Dijk, T.A. (1993). Principles of critical discourse analysis. *Discourse and Society Journal,* 4(2), 249-283.

Verheul, H. (1999). How social networks influence the dissemination of cleaner technologies to SMEs. *Journal of Cleaner Production,* 7, 213-219.

Verschuren, P., & Doorewaard, H. (1999). *Designing a research project.* LEMMA: Utrecht, the Netherlands.

Villarán F. (2004). Foro: responsabilidad social de la empresa. Perú2021 – COMEXPerú. Planeamiento Estratégico con Responsabilidad Social. CEPLAN.

Villarán, F. (2001). *Las PYMEs en la estructura empresarial peruana.* SASE: Peru.

Vives, A., Corral, A., & Isusi, I. (2005). *Responsabilidad social de la empresa en las PyMES de Latinoamérica.* Inter-American Development Bank and IKEI.

Vogl, C., Kilcher, L., & Schmidt, H. (2005). Are standards and regulations of organic farming moving away from small farmers' knowledge? *Journal of Sustainable Agriculture,* 26(1), 5-26.

Vonortas, N. (2002). Building competitive firms: technology policy initiatives in Latin America. *Technology in Society Journal,* 24, 433-459.

Wallace, T., Bornstein, L., & Chapman, J. (2007). *The aid chain. Coercion and commitment in development NGOs.* Practical Action Publication: Rugby, UK.

Wasserman, S., & Faust, K. (1994). *Social network analysis: Methods and applications.* Cambridge University Press: Cambridge, UK.

Wattanapinyo, A. (2006). *Sustainability of small and medium-sized agro-industries in Northern Thailand.* Thesis Wageningen University: Wageningen, the Netherlands.

Wellman, B. (1983). Network analysis. *Sociological Theory Journal,* 1, 155-200.

Wellman, B. (1988). Structural analysis: from method and metaphor to theory and substance. In: B. Wellman and S.D. Berkowitz (eds.) *Social Structures: A Network Approach.* Cambridge University Press: Cambridge, UK, pp. 19-61.

Wezel. A., Bellon, S., Dore´, T., Francis. C., Vallod, D., & David, C. (2009). Agroecology as a science, a movement and a practice. A review. *Agronomy for Sustainable Development Journal*. DOI: http://dx.doi.org/10.1051/agro/2009004.

Willer, H., & Kilcher, L. (eds.) (2009). *The world of organic agriculture – statistics and emerging trends 2009*. IFOAM: Bonn, Germany; FiBL, Frick; ICT: Geneva, Switzerland.

Willer, H., & Kilcher, L. (eds.) (2010). *The world of organic agriculture – statistics and emerging trends 2010*. IFOAM: Bonn, Germany; FiBL, Frick; ICT: Geneva, Switzerland.

Wood, Ch., & Roberts, B. (eds.) (2005). *Rethinking development in Latin America*. Pennsylvania State University Press: University Park, PA, USA.

World Bank (2006). Environmental sustainability: A key to poverty reduction in Peru. Country environmental analysis. Vol. 2: Full Report. Environmentally and Socially Sustainable Development Department Latin America and the Caribbean Region, WB. Document of the World Bank.

Wu, S., & Ansión, N. (2002). *BioMercado Perú, oferta y demanda de productos ecológicos*. Grupo Ecológica Perú e IFOAM. Lima, Perú.

Yussefi, M., & Willer, H. (2003). *The world of organic agriculture 2003. Statistics and future prospects*. IFOAM Publication, 5th revised edition, February 2003, IFOAM: Tholey-Theley, Germany.

Zelada, F. (2008). *Access to markets for small producers: 16 experiences from a model to assemble*. CORDAID. Mercadeando S.A.: Miroflores, Peru.

Zelenika, I., & Pearce, J.M. (2011). Barriers to appropriate technology growth in sustainable development. *Journal of Sustainable Development*, 4(6), 12-22.

Zucchetti, A., & Alegre, M. (2001). *Microempresa y ambiente*. ECOLAB SRL. OACA. Lima, Perú.

Zwetsloot, G.I.J.M., & Geyer, A. (1996). The essential elements for successful cleaner production programmes. *Journal of Cleaner Production*, 4(1), 29-39.

Websites consulted

Agencia Suiza para el Desarrollo y la Cooperacion, Secretaria de Estado para Asuntos Economicos (COSUDE/SECO)
http://www.cooperacion-suiza.admin.ch/peru (Accessed on 10 December 2011).

Asociación Peruana de Gastronomía (APEGA)
http://www.apega.pe (Accessed on 4 October 2010).

Asociación de Exportadores del Perú (ADEX)
http://www.adexperu.org.pe (Accessed on 10 December 2011).

Area Minera
http://www.aminera.com/index.php?option=com_content&id=11853 (Accessed on 24 January 2012).

BSR The Business of a Better World (BSA)
http://www.bsr.org/.

Catholic Organisation for Relief and Development Aid (CORDAID)
http://www.cordaid.nl (Accessed on 10 December 2011).

Centro de Asesoría Laboral del Perú (CEDAL)
http://www.cedal.org.pe (Accessed on 24 December 2011).

Centro de Ecoefiencia y Responsabilidad Social (CER)
www.cer.org.pe (Accessed on 24 October 2010).

Centro de Estudios para el Desarrollo y la Participación (CEDEP)
http://www.cedepperu.org (Accessed on 8 December 2011).

Crece MYPE La web del empresario MYPE

    http://www.crecemype.pe/portal/index.php/servicio-de-desarrollo-empresarial/premio-presidente (Accessed on 5 January 2012).

Cleaner production

    http://www.cleanerproduction.com (Accessed on 20 October 2010).

Climate Investment Funds (CIF)

    https://www.climateinvestmentfunds.org/cif/node/2 (Accessed on 26 January 2012).

Dutch interchurch organization for development cooperation (ICCO)

    http://www.icco.nl (Accessed on 9 December 2011).

Dutch interchurch organization for development cooperation (ICCO) and Kerk in Actie (KIA)

    http://www.iccokia.org/southamerica/ (Accessed on 15 January 2012).

European Office of Crafts, Trades and Small and Medium sized Enterprises for Standardisation (NORMAPME)

    http://www.normapme.eu/es/page/541/iso-26000-guia-de-aplicacion-normapme-para-pymes-europeas-disponible (Accessed on 20 January 2012).

Fondo de Investigacion y Desarrollo para la Competitividad FIDECOM (Innovate Peru)

    http://www.innovateperu.pe (Accessed on 8 December 2011).

Forum Empresa

    http://www.empresa.org (Accessed on 8 December 2011).

Grupo de Emprendimiento Ambiental (GEA)

    http://www.grupogea.org.pe (Accessed on 10 December 2011).

Innovación y Competividad para el Agro Peruano (INCAGRO)

    http://www.incagro.gob.pe (Accessed on 7 December 2011).

Instituto de Trasferencia de Tecnologías Apropiadas para Sectores Marginales (ITACAB)

    http://www.itacab.org (Accessed on 6 December 2011).

Inter-American Development Bank (IDB)

    http://www.iadb.org (Accessed on 14 October 2010).

International Federation of Organic Agriculture Movements (IFOAM)

    http://www.ifoam.org (Accessed on 10 November 2011).

International Finance Corporation (IFC)

    http://www.ifc.org/ifcext/climatechange.nsf/Content/CleanTechnologies (Accessed on 24 January 2012).

International Institute for Sustainable Development (IISD)

    http://www.iisd.org/business/tools/bt_eco_eff.aspx (Accessed on 15 January 2012).

International Network for Economic, Social and Cultural Rights (ESCR-Net)

    http://www.escr-net.org (Accessed on 4 October 2010).

International Organization for Standarization (ISO)

    http://www.iso.org (Accessed on 14 December 2011).

Ministerio de la Producción de Perú (PRODUCE)

    http://www.produce.gob.pe (Accessed on 5 December 2011).

Multilateral Investment Fund (MIF)

    http://www.iadb.org/mif/ (Accessed on 25 January 2012).

NESST

    http://www.nesst.org/peru/ramp/ (Accessed on 20 January 2012).

Oficina del Alto Comisionado para los Derechos Humanos (OHCHR)

    www.ohchr.org (Accessed on 5 October 2010).

Oxfam-Novib

    http://www.oxfamnovib.nl (Accessed on 14 December 2011).

PeruCompite

    http://www.perucompite.gob.pe (Accessed on 10 December 2011).

Programa Latioamericano de Responsabilidad Social Empresarial (PLARSE)

    http://www.plarse.org (Accessed on 14 December 2011).

Peru 2021

    http://www.peru2021.org (Accessed on 6 October 2010).

Programa Iberoamericano de Ciencia y Tecnología para el Desarrollo (Programa CYTED)

    http://www.cyted.org (Accessed on 4 October 2010).

Promoción del Desarrollo Sostenible (IPES)

    http://www.ipes.org (Accessed on 5 January 2012).

Red de Información en Consumo y Producción Sostenibles para América Latina y el Caribe (REDPYCS)

    http://www.redpycs.net (Accessed on 4 October 2010).

Red Latinoamericana de Producción más Limpia (CPLatinNet)

    http://www.produccionmaslimpia-la.net (Accessed on 10 December 2011).

Red Puentes Internacional

    http://www.redpuentes.org (Accessed on 14 December 2011).

Responsabilidad Social Comité Perú (iso2600peru)

    http://www.iso26000peru.org (Accessed on 4 October 2010).

Soluciones Prácticas

    http://www.solucionespracticas.org.pe (Accessed on 20 December 2011).

The National Center for Appropriate Technology

    http://www.ncat.org (Accessed on 10 December 2011).

The UN Global Compact

    http://www.unglobalcompact.org/case_story/378 (Accessed on 10 November 2011).

United Nations (UN)

    http://www.un.org/en/ (Accessed on 5 October 2010).

United Nations Environmental Programme (UNEP)

    http://www.unep.org (Accessed on 5 December 2011).

United Nations Industrial Development organization (UNIDO)

    http://www.unido.org (Accessed on 6 December 2011).

World Trade Organization (WTO)

    http://www.wto.org (Accessed on 6 October 2010).

World Bank

    http://www.worldbank.org (Accessed on 10 November 2011).

2020 Science Technology innovation in the 21st century

    http://2020science.org/category/technology-innovation-in-the-21st-century/ (Accessed on 25 January 2012).

Appendices

Appendix 1. List of interviews

Person interviewed	Organization/institution	Date
Alfonso Pablo Huerta Fernando, Director, ciencia and technology program	CONCYTEC	January 2011
Alfredo Oliveros, Head of the Clean Technology Programme	CONCYTEC	May 2006
Bruno Carpio, Project assistant	Peru 2021	December 2010
Carola Amezaga, National coordinator of the APOMIPE programme	MINKA-INTERCOOPERATION	April 2006
Cecilia Pardo, Head of the coordination board	Grupo Ecologica Peru	February 2008 March 2008
Cecilia Rizo-Patron, Project manager	Peru 2021	April 2006 August 2007 March 2008
César Paz, National representative, Agronomes et vétérinaires sans frontiers (AVSF), CICDA-Peru. Member affiliated to CEPICAFE	CICDA-Peru AVSF CEPICAFE	April 2008
Dora Cortijo Herrera, Manager	CET Perú	April 2006
Elizabeth Minaya, BID-CEDEP project manager	CEDEP	August 2007 April 2008
Ercilio Moura, Project manager of business social responsibility for SMEs	CEDAL	March 2008 November 2010
Fernando Alvarado, Head of the coordination board	Centro IDEAS	January 2008 March 2008 April 2009
Fernando Alvarado, Head of the coordination board	Grupo Ecologica Peru	November 2010
Frank Mechielsen, Coordinator of Red Puentes, Programma Officer, OXFAM-NOVIB, the Netherlands	OXFAM-NOVIB	October 2008
Gerdien Seegers, Programma Officer, CORDAID, the Netherlands	CORDAID	October 2008
Ines Carazo, Director of OTCIT	CITEs	May 2006
Janine Kuriger, Co-director	Swiss International Cooperation Agency	April 2006
Joke Langbroek, Business Development Advisor, ICCO, the Netherlands	ICCO	October 2008

Person interviewed	Organization/institution	Date
Jon Bickel, Head of PRAL project, SwissContact –Programa Regional de Aire Limpio	Swiss Contact	May 2006
Jose Sanchez Narvaez, Coordinator of science and technology	ITACAB	May 2009
Manual Hohagen Mory, Staff, Business development services	COPEME	May 2006
Marcos Alegre, Director	CER	January 2011
Maria Luisa Espinosa, Manager	CER	April 2008
Norma Vidal, Coordinator	ITACAB	May 2006
Pim Verhallen, Policy Advisor, ICCO, the Netherlands	ICCO	October 2008
Rosa Salas, Head of the National Programme of Cleaner Production	CONAM	May 2006
Silverio Trejo, Head of the coordination board	ANPE	March 2008 May 2009 November 2010
Silvia Wu, Executive director	RAE Peru	January 2008 March 2008 November 2010
Susanne Engelhardt, Programme Officer Latin-America, OXFAM-NOVIB, the Netherlands	OXFAM-NOVIB	October 2008
Victor Luque Luque, Coordinator of commercial affairs, Association of promotion and development	El Taller	November 2007

Appendix 2. Organic production networks: actors, platforms and connections

2a. Key Peruvian agro-ecological NGOs and their connections with key national, regional and global platforms.

Network of NGOs	Platforms		
	Peru	**Latin America**	**Global**
RAE Peru	Peruvian Agro-ecological Consortium, CONAPO Platform 'Peru, country free of transgenics', the Peruvian Network of Fair Trade and Ethical Consumption, and SNA	MAELA, GALCI	IFOAM
Grupo Ecologica Peru	Platform 'Peru, country free of transgenics'	MAELA, GALCI	IFOAM
ANPE	Peruvian Agro-ecological Consortium Platform 'Peru, country free of transgenics'	MAELA	Slow Food

Source: Alvarado, 2008, interview; Pardo, 2008, interview; Trejo, 2008, interview; Trejo, 2009, interview; Wu, 2008, interview.

2b. Description of national, regional and global platforms of agro-ecological NGOs.

National platforms of agro-ecological NGOs in Peru

CONAPO

The National Commission of Organic Products (CONAPO) was the coordination board of organizations in organic production to promote organic agriculture and propose regulations, norms and strategies on organic production to the Peruvian government. Funded by COSUDE and FAO, it has been established in 1998 and institutionalized in 2001. Initially CONAPO has been integrated by seven governmental entities such as the Ministry of Agriculture (MINAG); INRENA, INIA, SENASA, the Peruvian Export Promotion Agency (PROMPEX), the National Institute for the Defense of Competition and Protection of Intellectual Property (INDECOPI) and UNALM, four civil society organizations representatives such as RAE Peru, ANPE, JNC and the National Union for Small Banana Producers, and by one international cooperation agency representative such as the Swiss Agency for Development and Cooperation (COSUDE). CONAPO elaborated proposals for the government to build up the legal infrastructure (norms, institutions, etc.) within the Peruvian State. At the beginning, larger scale producer organizations such as JNC and the Union of Coffee Exporters have not been part of CONAPO.

In 2008 the Peruvian government set up the National Council of Organic Production (keeping the acronym CONAPO). After the approval of the national law of organic production in 2010, the former CONAPO was canceled as it reached its aim. Now, this new CONAPO is affiliated to MINAG as advisory agency in policies for organic production development. This new CONAPO will be composed by representatives from MINAG, Ministry of Production (PRODUCE), Ministry of Foreign Trade and Tourism (MINCETUR), INDECOPI, Regional Council of organic products (COREPOs), producers and NGOs. Up to December 2012, the new CONAPO is not operative yet.

Platform 'Peru, country free of transgenics'

The platform 'Peru, country free of transgenics' is integrated among others by RAAA, RAE Peru, ANPE, the Grupo Ecologica Peru, UNALM, CCE and ASPEC. Founded in 2007, it has become very active in lobbying against the importation to Peru of (genetically modified organisms (GMOs) based food. Initially, the platform was launched in 2006 with the name of Agro-ecological Consortium. Then, it evolved in the platform.

Peruvian Network of Fair Trade and Ethical Consumption

The Peruvian Network of Fair Trade and Ethical Consumption is integrated by the Economy of Solidarity Network in Peru (GRESP), the National Union for Coffee Producers (JNC), the Ecological Agriculture Network of Peru (RAE Peru), the Child and Adolescent Workers' Association (MANTHOC), Life Promotion (FOVIDA), Episcopal Commission for Social Action (CEAS), Organization of Peruvian artisans of Peru (CIAP) and Caritas del Peru.

Peruvian Agro-ecological Consortium

PAC has been established in 2009 and it is the formal coordination board of PAM. The board members are ANPE, RAE Peru, RAAA and ASPEC. At PAM, NGOs and producer associations coordinate division of tasks among PAM members and decisions about join projects and policy proposals in order to perform better as movement and negotiate in better position with the Peruvian state. The Consortium gets support from the Aid for Development Gembloux (ADG).

Peruvian Agro-ecological Movement

PAM is a wider non-formal network of NGOs, producers, small-scale agri-enterprises, universities, consumers and other civil society organizations coping with organic production and commercialization of organic food in Peru. The main representatives are RAE Peru, RAAA and ANPE.

National Environmental Society

SNA is an umbrella network of environmental NGOs. It groups to 27 organizations; including networks of NGOs and individual NGOs working on environmental protection in Peru by promoting dialogue and improvement of environmental conditions in order to benefit the population. Key NGOs of PAM belong to SNA. SNA's activities include analyzing environmental policy, resolving environmental conflicts, promoting the participation of citizens and organizing environmental discussions at the national level.

Regional platforms of agro-ecological NGOs in Latin America

Latin American agro-ecological movement

MAELA is integrated by 150 organizations affiliated in 20 countries such as NGOs, universities, producers, local networks and grass root organizations aimed to strengthening ecological agriculture towards sustainable development. MAELA has been founded in 1992.

IFOAM Regional Group for Latin America and the Caribbean

GALCI is the regional branch of IFOAM, where belong producer associations, traders, certifiers, NGOs and individuals from Latin America and the Caribbean. GALCI has a high level capacity to lobbying for organic agriculture, integrated by IFOAM members and supporters and individuals working in favor of organic agriculture in the region. GALCI's aim among other aims are fomenting action concerning IFOAM policies, investigating local, regional and national development of organic markets, serving as a forum for the discussion and analysis of the contemporary situation for organic agriculture in Latin America, the Caribbean and the world.

>>

Global platforms of agro-ecological NGOs

International Federation of Organic Agriculture Movements
   IFOAM is a global platform for the organic movement stakeholders – from farmers' organizations to
   multinational certification agencies, ensuring the credibility and longevity of organic agriculture as a
   means to ecological, economic and social sustainability.
   IFOAM's mission is leading, uniting and assisting the organic movement in its full diversity. IFOAM
   promotes the development of organic markets.

Slow Food Movement / Slow Food Foundation
   Slow Food is a non-profit, eco-gastronomic member-supported organization that was founded in 1989
   to counteract fast food and fast life, the disappearance of local food traditions and people's dwindling
   interest in the food they eat where it comes from, how it tastes and how our food choices affect the rest
   of the world. Slow Food network has over 100,000 members in 132 countries.
   Slow Food Italy: The Slow Food Foundation for Biodiversity is part of Slow Food and was founded
   in Italy in 2003. The Slow Food Foundation supports projects in defense of food biodiversity in more
   than 50 countries and promotes a sustainable agriculture that respects the environment, the cultural
   identity of local people, and promotes animal well-being. On a local level, Slow Food Convivia brings
   producers and consumers closer together. In Peru, Convivium Lima is the local branch of Slow Food.
   In Peru Slow Food support agrobiodiversty projects of ANPE.

Source: Alvarado, 2008, interview; Pardo, 2008, interview; Trejo, 2008, interview; Trejo, 2009, interview;
Wu, 2008, interview; Institutional web sites.

2c. Key Peruvian agro-ecological NGOs and their international 'partner' NGOs.

Peruvian Network of NGOs	International NGOs
RAE Peru	Interchurch organization for development cooperation (ICCO)
	Heifer International
	Food and Agriculture Organization of the United Nations (FAO)
	German Development Service (DED)
	Bread for the World
	The German catholic bishops' organization for development cooperation (MISEREOR)
Grupo Ecologica Peru	Catholic Organization for Relief and Development AID (CORDAID)
	OXFAM-NOVIB
	German Development Service (DED)
	German Public-private-partnership projects (PPP projects)
	Schmitz Foundation
	German catholic bishops' organization for development cooperation MISEREOR
ANPE	Humanist Institute for Cooperation with Developing Countries (HIVOS)
	Bread for the World
	the Aid for Development Gembloux (ADG)
	Italian Municipalities

Source: Alvarado 2008, interview; Pardo, 2008, interview; Trejo, 2008, interview; Wu, 2008, interview.

2d. Other NGOs in Peru that support agri-SMEs development.

The Consortium of Private Organizations for the Promotion of Small and Medium Business Development of Peru (COPEME)

COPEME is a network of about 60 NGOs among them the Centro IDEAS and el Taller, supporting SMEs development in rural and urban areas. The network works for further formalization, competitiveness and productivity of SMEs meanwhile strengthening local economies. COPEME gets funding mainly from HIVOS, ICCO, OXFAM and MISEREOR.

Peruvian Association of Consumers and Users (ASPEC)

ASPEC raises consumers' awareness of good public and private services provision in health, food, transport, education and other relevant matters for consumers. ASPEC act as a watch dog for the proper application of norms and regulation in market. ASPEC works closely with CCE. For proper food consumption, ASPEC organizes awareness campaigns on GMOs and food labeling.

National Agrarian Confederation (CNA)

CNA is the union of small scale producers of Peru. It represents 16 regional committees throughout Peru. CNA aims to sustainable development of small scale agriculture. Mayor funders: AGRITERRA, OXFAM-NOVIB, Bread for the World and international farmers unions.

Peasant Confederation of Peru (CCP)

CCP groups to 19 federations of peasants, small scale producers, community enterprises and grass root organizations throughout Peru aimed to stand up for human rights, democracy, gender equality and rural development of Peru.

National Users Council of the Irrigation Districts of Peru (JNUDRP)

JNUDRP groups 112 districts of irrigation throughout Peru. JNUDRP is a network that represents to users of water for irrigation purposes, protects their producer affiliates and proposes norms and regulations to the Peruvian government.

National Union for Coffee Producers (JNC)

JNC is a network of 34 coffee cooperatives, grouping 38000 family affiliates from 14 coffee plantation areas from Peru. JNC's affiliates include organic, fair trade and other social and environmental friendly coffee producers. JNC represents to Peruvian coffee producers. JNC offers services to their producer affiliates, capacity building in business management, market access and links to international fairs.

Program for Exchange, Dialogue and Consultation on Sustainable Agriculture and Food Sovereignty (PIDAASSA PERU)

PIDAASSA is the coordination board of NGOs at local, national and Latin American level aimed to dialogue, exchange of experiences and participatory learning on sustainable agriculture and food security and the methodology peasant to peasant. In Peru, PIDASSA group 11 Peruvian NGOs among them IDMA, ANPE, PIDECAFE and CNA. At regional level, the CNA group 76 NGOs both local and international from 11 countries of Latin America, the most of them partners of the international cooperation agency Bread for the World. PIDAASSA gets financial support from Bread for the World.

National Convention for Peruvian Agriculture (CONVEAGRO)

CONVEAGRO is a forum of organizations to analyze, evaluate and coordinate the agrarian situation of Peru. The Forum groups to 32 organizations including large and medium sized producer unions, agrarian professionals, NGOs, research institutes and agrarian media.

Agrarian Coffee Cooperative of Cuzco (COCLA)

COCLA is an enterprise of coffee cooperatives. It provides technological and expertise support for coffee production. COCLA group to 7,500 small scale producers, clustered in 23 cooperatives of first floor, covering a production area of 21,000 hectares.

Integrated Program for Coffee Development (PIDECAFE)

PIDECAFE promotes capacity building, implementation of environmental friendly farm practices and income opportunities base on better access to markets and the proper access food for producer families. The intervention approach integrates productive, economic environmental, social and political sides of the production in rural areas. PIDECAFE works with 10 producer cooperatives and associations, especially with the producers association CEPICAFE and 2 municipalities in the northern part of Peru.

Institute for Small Sustainable Production (IPPS)

IPPS has been founded in 2001 in Lima, Peru, to address social, environmental, and economic problems of small-scale farmers and to increase their capacity. The institute also emphasizes interdisciplinary research and innovation in the development of rural communities.

Source: Institutional websites.

2e. Peruvian governmental agencies that support agri-SMEs development.

Ministry of Agriculture's General Direction of Agrarian Competitiveness (DGCA)

DGPA promotes and links producers and rural enterprises with resources and services from public agencies, NGOs and business in order to improve their competiveness and commercialization for exportation.

National Service of Agrarian Health (SENASA)

SENASA is the national authority in agrarian health, controlling seeds quality, organic production and safety of national food production.

SENASA provides official guarantee of organic products for international markets.

National Institute for Agrarian Innovation (INIA)

INIA generates knowledge and technology innovation as an answer to market demands, and the transfers of them to local producers. INIA develops technologies of organic production.

Cooperation Fund for Social Development (FONCODES)

FONCODES finances small projects of social and economical infrastructure, and capacity building for production in rural areas oriented to local and international market. FONCODES works in alliance with municipalities and NGOs.

Presidency of the Ministries' Council (PCM)

PCM is a governmental agency that promotes, coordinates and links the overall governmental policy between the executive power and other agencies from the Peruvian state.

Source: Trejo, 2009, interview; Institutional web sites.

Appendix 3. Interview guidelines for local NGOs, international NGOs and SMEs in the domains of organic production, business social responsibility and sustainable production

3a. Organic production: interview guideline for local NGOs.

Issue	Items	Questionnaire
I. General issues		
General issues		
General Profile of the Organization	Characteristics Relevance of environment and SMEs in organization policies	What projects is the NGO implementing towards sustainability of SMEs? What is the geographical scope of the intervention (local, national, etc.)? Specify the names of those geographical areas.
		How central are environment and SMEs in organization policies (statutes set of principles, etc.)? What are the key features of the beneficiaries (SMEs)?
Motivation and constrains	Motivation of support Constrains to overcome	What has been the motivation to support SMEs (opportunities of funding, organizational commitment, etc.)? What are the forces working on the NGO to look at SMEs (market pressure, local motivations, SMEs growth, etc.)?
		What are the limiting factors facing NGOs for fostering sustainable SMEs?
II. Network mapping		
Network relationships		
Actors	Network members and partners	Members and other partner organizations (local, national and international) that support SMEs.
Connections	Coordinating mechanism Collaboration characteristics, schemes and constrains	What type of network relationship (e.g. participatory, arbitrary, top-down, bottom-up, etc.) is applied among local NGO, international NGO and other network partners to deliver support towards SMEs?
		How the network members coordinate their efforts to accomplish their tasks? What coordinating mechanism (e.g. mutual adjustment, direct supervision and standardization) is most applied?
		How structure is the network ties (e.g. formally work as an organization, informal network, etc.)?
		How NGOs and SMEs collaborate? Are the SMEs organized in any association or do they work individually? What are the motivations of SMEs to work associatively? What are the critical barriers to work collaboratively and how can they be overcome?

Issue	Items	Questionnaire
Nodes and power	Key and influential organizations in the Network	If you want to achieve something, which of the partner organizations do you need most? Why? Who sets the agenda and policies in supporting sustainability of SMEs?
Resource exchanged	Exchange of resources	What is exchanged (money, knowledge, etc.) in network relationship and in the support? How the NGOs is facilitating the better performance of SMEs in market and the access to new markets in terms of access to capital, technology, knowledge, etc.? How do the NGO and SMEs benefit from the exchange? Do the SMEs get satisfied?
Main challenges	Main changes in the past and challenges in future for NGOs, their Network and SMEs	What has changed most in the NGOs/network structure in the last 10 years (size, number of network members, alliances, network consolidation, etc.)?
		What organizations or entities are seen as potential partners, besides SMEs and consumers? What challenges do the NGO and their network see in the near future in delivering their support for SMEs? Is the NGO capable to take that challenge?
		What challenges do the NGO and their network see in the near future in delivering their support for SMEs? Is the NGO capable to take that challenge? What weakness is seen to further the movement (e.g. the community organization, added value, business skills, etc.)
		In which ways do SMEs representatives participate in the network? Are they capable to propose initiatives? What is constraining for more active participation of SMEs? What want to achieve the SMEs in future?
		How the voice and concerns of SMEs and local NGOs reach the larger regional and global networks (MAELA, IFOAM, Fair trade, etc.)? Are there proper channels?
		How those larger global networks of organic production influence global and regional politics (e.g. UN, WTO, CAN, TLC, etc.)?

Issue	Items	Questionnaire
III. Discourse mapping		
Framing		
Framing	Mission, vision and objectives	What are the institutional aims (mission, vision and objectives) of the NGO?
	Sustainability of SMEs	How the NGO understand 'sustainability' of SMEs? Which
	Definition of OP	aspect of sustainability do you think is more important
	NGOs and market	(economic, social, ecological, etc.)?
	NGOs and SMEs development	How the NGO define OP and how the organization has achieved this definition (policy of the network, guidelines of
	Economic vs. social role of NGOs	funding agencies, etc.)?
	Global concerns and SMEs needs	To what extent the NGO is along with free-market as a way to organize economic relationships? If the NGOs is in line, how is the NGOs dealing with environmental concerns related to mass-production & consumption? If the NGO is opposing/ resisting to global forces how they are cooping with the need for SMEs economic pull-out?
		Does the NGO prioritize the support of SMEs producing for local or international market?
		How does the NGO harmonize their dual 'economic' and 'social' role? Does the NGO consider closer to business or closer to the state?
		How OP can help to harmonize global concerns (governance, climate change, etc.) with local SMEs needs (technology, capital, market, etc.)? How NGOs are positioning their discourse in national, regional and global politics?
Reframing		
Change and future trends	Changes in thinking and identity	What has changed most in the NGOs/network in the last 10 years (organization's views, insights, movement thoughts, limitations in the discourse for wide-spreading, etc.)?
	New views and learnings	What lessons have been learned working with SMEs? How are those learnings incorporated in recent projects to support SMEs?
		What tasks would the NGO like to perform towards SMEs in the future and how it would differ from the actual role?
		Has the NGO already established an intervention plan for future? If so, how does look like this plan? What new projects and/or programs the network is planning to implement towards SMEs?

3b. Business social responsibility: interview guideline for local NGOs.

Issue	Items	Questionnaire
I. General issues		
General issues		
General profile of the organization	Characteristics Relevance of environment and SMEs in organization policies	What projects is the NGO implementing towards sustainability of SMEs? What is the geographical scope of the intervention (local, national, etc.)? Specify the names of those geographical areas. How central are environment and SMEs in organization policies (statutes set of principles, etc.)? What are the key features of the beneficiaries (SMEs)?
Motivation and constrains	Motivation of support Constrains to overcome	What has been the motivation to support SMEs (opportunities of funding, organizational commitment, etc.)? What are the forces working on the NGO to look at SMEs (market pressure, local motivations, SMEs growth, etc.)? What are the limiting factors facing NGOs for fostering sustainable SMEs?
II. Network mapping		
Network relationships		
Actors	Network members and partners	Members and other partner organizations (local, national and international) that support SMEs.
Connections	Coordinating mechanism Collaboration characteristics, schemes and constrains	What type of network relationship (e.g. participatory, arbitrary, top-down, bottom-up, etc.) is applied among local NGO, international NGO and other network partners to deliver support towards SMEs? How the network members coordinate their efforts to accomplish their tasks? What coordinating mechanism (e.g. mutual adjustment, direct supervision and standardization) is most applied? How structure is the network ties (e.g. formally work as an organization, informal network, etc.)? How NGOs and SMEs collaborate? Are the SMEs organized in any association or do they work individually? What are the motivations of SMEs to work associatively? What are the critical barriers to work collaboratively and how can they be overcome?

Issue	Items	Questionnaire
Nodes and power	Key and influential organizations in the Network	If you want to achieve something, which of the partner organizations do you need most? Why? Who sets the agenda and policies in supporting sustainability of SMEs?
Resource exchanged	Exchange of resources	What is exchanged (money, knowledge, etc.) in network relationship and in the support? How the NGOs is facilitating the better performance of SMEs in market and the access to new markets in terms of access to capital, technology, knowledge, etc.? How does the NGO and SMEs benefit from the exchange? Do the SMEs get satisfied?
Main challenges	Main challenges for NGOs, their Network and SMEs	What has changed most in the NGOs/network structure in the last 10 years (size, number of network members, alliances, network consolidation, etc.)?
		What organizations or entities are seen as potential partners, besides SMEs and consumers?
		What challenges do the NGO and their network see in the near future in delivering their support for SMEs? Is the NGO capable to take that challenge? What weakness is seen to further the movement (e.g. the community organization, added value, business skills, etc.)?
		In which ways do SMEs representatives participate in the network? Are they capable to propose initiatives? What is constraining for more active participation of SMEs? What want to achieve the SMEs in future?
		How the voice and concerns of SMEs and local NGOs can be taken in account in the larger global networks (WBCSD, Forum Empresa, Global Compact, etc.)?
		How those larger global networks of BSR influence global and regional politics (e.g. UN, WTO, CAN, TLC, etc.)?

Issue	Items	Questionnaire
III. Discourse mapping		
Framing		
Framing	Mission, vision and objectives	What are the institutional aims of the NGO?
		How the NGO understand 'sustainability' of SMEs? Which
	Sustainability of SMEs	aspect of sustainability do you think is more important
	Definition of BSR	(economic, social, ecological, etc.)?
	NGOs and market	How the NGO define BSR and how the organization has
	NGOs and SMEs development	achieved this definition (policy of the network, guidelines of funding agencies, etc.)?
	Economic vs. social role of NGOs	To what extent the NGO is along with free-market as a way to organize economic relationships? If the NGOs are in line,
	Global concerns and SMEs needs	how are the NGOs dealing with environmental concerns related to mass-production & consumption? If the NGO is opposing/resisting to global forces how they are cooping with the need for SMEs economic pull-out?
		Does the NGO prioritize the support of SMEs producing for local or international market?
		How does the NGO harmonize their dual 'economic' and 'social' role? Does the NGO consider closer to business or closer to the state?
		How BSR can help to harmonize global concerns (governance, climate change, etc.) with local SMEs needs (technology, capital, market, etc.)? How NGOs are positioning their discourse in national, regional and global politics?
Reframing		
Change and future trends	Changes in thinking and identity	What has changed most in the NGOs/network in the last 10 years (organization's views, insights, movement thoughts,
	New views and learnings	limitations in the discourse for wide-spreading, etc.)?
		What lessons have been learned working with SMEs? How are those learnings incorporated in recent projects to support SMEs?
		What tasks would the NGO like to perform towards SMEs in the future and how it would differ from the actual role?
		Has the NGO already established an intervention plan for future? If so, how does look like this plan? What new projects and/or programs the network is planning to implement towards SMEs?

3c. Sustainable production: interview guideline for local NGOs.

Issue	Items	Questionnaire
I. General issues		
General issues		
General profile of the organization	Characteristics Relevance of environment and SMEs in organization policies	What projects is the NGO implementing towards sustainability of SMEs? What is the geographical scope of the intervention (local, national, etc.)? Specify the names of those geographical areas. How central are environment and SMEs in organization policies (statutes set of principles, etc.)? What are the key features of the beneficiaries (SMEs)?
Motivation and constrains	Motivation of support Constrains to overcome	What has been the motivation to support SMEs (opportunities of funding, organizational commitment, etc.)? What are the forces working on the NGO to look at SMEs (market pressure, local motivations, SMEs growth, etc.)? What are the limiting factors facing NGOs for fostering sustainable SMEs?
II. Network mapping		
Network relationships		
Actors	Network members and partners	Members and partner organizations (local, national and international) that support SMEs.
Connections	Coordinating mechanism Collaboration characteristics, schemes and constrains	What type of network relationship (e.g. participatory, arbitrary, top-down, bottom-up, etc.) is applied among local NGO, international NGO and other network partners to deliver support towards SMEs? How the network members coordinate their efforts to accomplish their tasks? What coordinating mechanism (e.g. mutual adjustment, direct supervision and standardization) is most applied? How structure is the network ties (e.g. formally work as an organization, informal network, etc.)? How NGOs and SMEs collaborate? Are the SMEs organized in any association or do they work individually? What are the motivations of SMEs to work associatively? What are the critical barriers to work collaboratively and how can they be overcome?

Issue	Items	Questionnaire
Nodes and power	Key and influential organizations in the Network	If you want to achieve something, which of the partner organizations do you need most? Why? Who sets the agenda and policies in supporting sustainability of SMEs?
Resource exchanged	Exchange of resources	What is exchanged (money, knowledge, etc.) in network relationship and in the support? How the NGOs is facilitating the better performance of SMEs in market and the access to new markets in terms of access to capital, technology, knowledge, etc.? How does the NGO and SMEs benefit from the exchange? Do the SMEs get satisfied?
Main challenges	Main challenges for NGOs, their Network and SMEs	What has changed most in the NGOs/network structure in the last 10 years (size, number of network members, alliances, network consolidation, etc.)?
		What organizations or entities are seen as potential partners, besides SMEs and consumers?
		What challenges do the NGO and their network see in the near future in delivering their support for SMEs? Is the NGO capable to take that challenge? What weakness is seen to further the movement (e.g. the community organization, product added value, business skills, etc.)?
		In which ways do SMEs representatives participate in the network? Are they capable to propose initiatives? What is constraining for more active participation of SMEs? What want to achieve the SMEs in future?
		How the voice and concerns of SMEs and local NGOs can be taken in account in the larger global networks (IDB/FOMIN, Red CAB-PL, UNIDO-NCPCs, etc.)?
		How those larger global networks of Cleaner Production influence global and regional politics (e.g. UN, WTO, CAN, TLCs, etc.)?

>>

Issue	Items	Questionnaire

III. Discourse mapping

Framing

Framing	Mission, vision and objectives Sustainability of SMEs Definition of SP NGOs and market NGOs and SMEs development Economic vs. social role of NGOs Global concerns and SMEs needs	What are the institutional aims of the NGO? How the NGO understand 'sustainability' of SMEs? Which aspect of sustainability do you think is more important (economic, social, ecological, etc.)? How the NGO define SP and how the organization has achieved this definition (policy of the network, guidelines of funding agencies, etc.)? To what extent the NGO is along with free-market as a way to organize economic relationships? If the NGOs are in line, how are the NGOs dealing with environmental concerns related to mass-production & consumption? If the NGO is opposing/resisting to global forces how they are cooping with the need for SMEs economic pull-out? Does the NGO prioritize the support of SMEs producing for local or international market? How does the NGO harmonize their dual 'economic' and 'social' role? Does the NGO consider closer to business or closer to the state? How SP can help to harmonize global concerns (governance, climate change, etc.) with local SMEs needs (technology, capital, market, etc.)? How the NGO is positioning their discourse in national, regional and global politics?

Reframing

Change and future trends	Changes in thinking and identity New views and learnings	What has changed most in the NGOs/network in the last 10 years (organization's views, insights, movement thoughts, limitations in the discourse for wide-spreading, etc.)? What lessons have been learned working with SMEs? How are those learnings incorporated in recent projects to support SMEs? What tasks would the NGO like to perform towards SMEs in the future and how it would differ from the actual role? Has the NGO already established an intervention plan for future? If so, how does look like this plan? What new projects and/or programs the network is planning to implement towards SMEs?

3d. Interview guideline for international NGOs.

Partners in Peru

What organizations and/or projects related to sustainability of enterprises does your organization support in Peru?

What are the major local partners?

In which ways does your organization support these local partners?

Does your organization have fixed partners in Peru or do they change in time? Is there any potential partner emerging currently in Peru?

What strategies does your organization use to ensure that the projects financed achieve their aims?

Partners in the north

Are there other key partners here in the Netherlands, or at the international level involved in the projects of the Entrepreneurship programme sector? What is their contribution to these projects?

Rationale

For the projects mentioned above, does your organization apply specific rationality or principles to guide the intervention?

What is the role of sustainability concept in the intervention approach and how is it operationalized?

Is there a focus on producers and small enterprises? If so, for what reasons is there focus on bringing farmers to market? What is expected for farmers from this moving to market?

Viewpoints

What are, in your view, the main barriers or difficulties to support producers to access and perform better in market?

Are there good examples of partners/projects in Peru that cover your expectation?

Why does your organization work mainly with civil society organizations in Peru to deliver support to producers and small agri-enterprises? Why not to work with large business or the government?

Are NGOs getting closer to business rationale? How does your organization harmonize the 'social' and 'economic' roles? Is there any conflict of this 'dual' role, considering the more central role of ecological concerns and the sustainability issue worldwide next to the widely expansion of market?

How does the feedbacks, if any, from local partners and beneficiaries in Peru, enrich your organization's views?

3e. Interview guideline for SMEs.

I. General issues

How central is environment in SME policies (set of principles, production, commercialization, etc.)?

What are the forces working on the SME to look at environment as a business strategy (regulation, market pressure, personal motivation, etc.)?

Is there any NGO playing a role in supporting projects on organic production, business social responsibility and sustainable production? Describe briefly the project (learnings, challenges). What are the key features of the NGO? What has been the motivation of the NGO to support the SME (opportunities of funding, organizational commitment, etc.)?

What are the limiting factors facing SME to include environmental dimension within SME?

II. Network mapping

What type of network relationship (e.g. participatory, arbitrary, top-down, bottom-up, etc.) is applied among SME/cluster and the NGO?

How the network/cluster members coordinate their efforts to accomplish their tasks? What coordinating mechanism (e.g. mutual adjustment, direct supervision and standardization) is most applied?

How structured is the network ties (e.g. formally work as an organization, informal network, etc.)?

How SMEs and NGOs collaborate? Are the SMEs organized in any association/cluster or they work individually? What are the motivations of SMEs to work associatively? What are the critical barriers to work collaboratively and how can they be overcome? Are there local, national and international partner organizations involved in the network/cluster?

If you want to achieve something, which of the partner organizations do you need most? Why? Who sets the agenda and policies in the network/cluster of SMEs?

What is exchanged (money, knowledge, etc.) in network relationship and in the support among SMEs and NGOs?

How the NGOs is facilitating (or how an organization might facilitate) the better performance of SMEs in market and the access to new markets in terms of access to capital, technology, knowledge, etc.? How does the NGO and SMEs benefit from the exchange? Do the SMEs get satisfied?

What has changed most in the SMEs/cluster structure in the last 10 years (size, number of network members, alliances, network consolidation, etc.)?

What organizations or entities are seen as potential partners, besides NGOs?

What challenges do the SMEs and their networks see in the near future in making business sustainable? Is the SME capable to take that challenge? What weakness is seen to further environment and sustainability in the business and markets (e.g. the community organization, added value, business skills, consumer awareness, etc.)?

>>

In which ways do SMEs representatives participate in the NGO's network/cluster? Are SMEs capable to propose initiatives? What is constraining for more active participation of SMEs? What want to achieve the SMEs in future?

How the voice and concerns of SMEs and local NGOs can be taken into account in the larger global networks (IFOAM, Fair trade, MAELA/IDB/FOMIN, Red CAB-PL, UNIDO-NCPCs/WBCSD, Forum Empresa, Global Compact, etc.)?

How those larger global networks of OP, CP and BSR influence global and regional politics (e.g. UN, WTO, CAN, TLC, etc.)?

III. Discourse mapping

What are the institutional aims of the SMEs?

How the SME understand 'sustainability'? Which aspect of sustainability do you think is more important (economic, social, ecological, etc.)?

How the SME define BSR, CP or OP and how the organization has achieved this definition (policy of the business, exporters, banks, funding agencies, government, etc.)?

Does the SME is along with free-market as a way to organize economic relationships? How is the SME dealing with environmental concerns related to his production & market? If the SME is an environmentally responsible business, how they are cooping with the need for economic growth and better performing in market?

Does the SME prioritize producing for local or international market? Why?

How do the SMEs harmonize their dual 'economic' and 'social' role? Do the SMEs consider itself as an example of 'sustainable business'?

How BSR, CP or OP can help to harmonize global concerns (governance, climate change, etc.) with local SMEs needs (technology, capital, market, etc.)? How SMEs and business networks are positioning the sustainable business discourse in national, regional and global politics?

What has changed most in the SME in the last 10 years (organization's views, insights, market, engagement with wider networks, etc.)?

If cooperation or support has been got from NGOs, what lessons have been learned working with them? How are those learnings incorporated in recent projects of business sustainability?

What tasks would like SME take up or strengthening and what tasks live out to NGOs or other partners?

Has the SMEs already established an intervention plan for future? If so, how does look like this plan? What new projects and/or programs the SMEs are planning to implement?

Summary

The importance of small and medium-sized enterprises (SMEs) in terms of employment and income generation has been recognized worldwide. In Peru, SMEs are responsible for 85% of the employment at the national level and they represent 98% of the total companies registered. Around 12% of SMEs, organized in associations, clusters, and cooperatives or as single companies, are dedicated to productive actives; the others are engaged in commercial and services activities. However, next to their positive economic role, SMEs are also responsible for significant disturbances of nature, environmental degradation and threats to human health. Environmental pollution related to the increase of productive activities has become evident in Peru and the entire region of Latin America.

The thesis aims to provide a better understanding of the changing roles of NGOs in promoting sustainability of SMEs in Peru, using the perspectives of networks and discourses. It focuses on three domains, which together are characteristic for promoting of SME sustainability in Peru: organic production (the first case study), business social responsibility (the second cases study) and sustainable production (the third case study). Three research questions have been outlined for this research: First, what are the networks of NGOs promoting sustainability of SMEs involved in the domains of organic production, business social responsibility and sustainable production in Peru, and what are the main changes in time in these networks? Second, what are the main discourses fostering sustainability that prevail and are articulated in these networks of NGOs and what are the main changes in time in these discourses? And finally, how to understand and assess the actual, new and potential roles of NGOs in promoting sustainability of SMEs in terms of network society theory and ecological modernization theory?

In this study the universe of NGOs is narrowed to NGOs operating in Peru that provide support (a) to medium and small scale producers and producer associations to bring organic products to local and global markets, (b) to urban and rural small scale enterprises to adopt cleaner production and appropriate technologies, and/or (c) to SMEs to upgrade social and environmental standards within value chains involving large companies. Some SMEs are concentrated in the main cities of Peru such as Trujillo, Arequipa and Lima, while other SMEs, such as organic food producers, are spread all over the country. In any case, SMEs under this research have collaboration ties with the NGOs to be studied. The research questions were investigated by means of more than 28 interviews with representatives of local NGOs, international NGOs, local SMEs and the national government, carried out in the period of 2006 to 2010. Additionally, documents and internet sources were consulted.

The networks involved in promoting the sustainability of SMEs are: the agro-ecological network, the organic market network and the ecological farming network in the first case study; the social justice network and the business network in the second case study; and the eco-efficiency network, the appropriate technology network, the cleaner technology network, the technological innovation network and the urban cleaner production network in the third case study.

The main actors identified in the networks of the organic production domain are: the Ecological Agriculture Network of Peru (RAE Peru), Grupo Ecologica Peru and the National Ecological Producers Association (ANPE). RAE Peru consists of 16 individual NGOs operating throughout Peru and has led several initiatives (e.g. Biocanastas, Bioferias, Biostores) to develop the organic market in Peru. Grupo Ecologica Peru consists of 5 NGOs and 24 producers, including associations and individual producers, and it commercializes organic products at local competitive markets (e.g. the Bioferia Miraflores farmers' market) and provides the supply of organic food to supermarkets. ANPE Peru consists of 22 organic small scale producer associations (including small food processers and family small-scale enterprises). ANPE's constituencies produce and commercialize organic food in 13 farmers' markets throughout the country.

The main actors identified in the business social responsibility networks are: the Labor Advisory Council of Peru (CEDAL), the Center of Studies for Development and Participation (CEDEP) and Peru 2021. CEDAL and CEDEP promote business social responsibility for urban and rural small enterprises in Peru in order to meet national regulation and international standards on labor rights and good environmental practices. CEDAL has been collaborating with 60 small enterprises of garment and handy craft makers, organized in clusters, who commercialize their products directly to consumers or business intermediaries oriented at domestic and foreign markets. CEDEP collaborates with small and medium-sized agri-industries, small garment workshops, shoemakers, metal workshops and bakeries to adopt business social responsibility principles by improving working conditions for their employees and sustainable production practices. Peru 2021 collaborates only with SMEs that are providers of larger companies in value chains promoting social and environmental standards.

The main actors involved in the sustainable production networks are: the Eco-efficiency and Social Responsibility Center (CER), the Institute for the Transfer of Technology for Marginal Sectors (ITACAB), the National Council of Science and Technology (CONCYTEC), the Centers of Technological Innovation (CITEs) and the Peruvian Institute of Social Economy (IPES). While CER, IPES and most of CITEs are NGOs, ITACAB and CONCYTEC are (inter)governmental agencies. CER provides consultancy for small scale suppliers of larger domestic companies and single SMEs exporting to international markets. Through the projects EcoADEX, EcoHotels and EcoParks CER aims to reduce greenhouse gas emissions, increase eco-efficiency, optimize production and services processes and reduce operation costs in SMEs. ITACAB promotes technological transfer to small scale rural enterprises through the Center for Technological Transferring Resources. CONCYTEC promotes technological transfer for SMEs but it is currently dispersed into several institutional programmes. CITEs provide production technologies services to SMEs. In total there are 13 CITEs throughout Peru, each one specialized in particular type of products (e.g. leather and shoemaking, wood and furniture, wine and horticulture, tropical fruits and medicinal plants, garment, agro-industry, textile, logistic and tracing, software and forest wood). Finally, IPES promotes cleaner technologies in small scale industries and workshops located in urban areas. During the last year, IPES is focusing on the establishing of recycling SMEs of electronic waste. As this overview shows, in the sustainable production domain, not only NGOs perform central roles but also governmental agencies. In some cases, quite close cooperation occurs between NGOs and governmental agencies.

In all three cases, networks of sustainability of SMEs are structured as interlinked platforms operating at local, national, Latin American and global level. For instance, in the organic production networks the Bioferias are at the local level, the Peruvian Agroecological Consortium at the national level, MAELA and GALCI at Latin American level and IFOAM at global level. Platforms include civil society, market and state actors. For instance, in the business social responsibility networks the civil society actors are CEDEP and CEDAL, in the organic production networks the market actors are the small scale enterprises affiliated with ANPE and Grupo Ecologica Peru, and in the sustainable production networks the state actors are CONCYTEC (governmental agency), ITACAB (inter-governmental agency) and the CITEs central office (OTCIT). Coordination and channeling of resources in the network platforms are performed by key actors, such as RAE Peru, Grupo Ecologica Peru, ANPE, Peru 2021, CEDAL, CEDEP, CER, CONCYTEC, ITACAB, OTCIT and IPES.

The ten networks are composed by diverse types of NGOs. Next to conventional NGOs as key actors, producer NGOs, market NGOs, business NGOs, technocratic NGOs and government organized NGOs (GONGOs) have emerged. Although NGOs are central in most networks, (inter)governmental agencies (GONGOs) are also central in the cleaner technology network, the appropriate technology network and the technological innovation network. CONCYTEC, ITACAB and the CITES' central office (OTCIT) are agencies that are part of the governmental structure, but they operate in practice pretty much as NGOs. Hence, NGOs and these (inter) governmental agencies perform similar roles in the networks, compete for funding and operate projects funded by international cooperation agencies. Therefore, the (inter)governmental agencies (GONGOs) that are part of these networks of sustainability of SMEs has been found out to be less effective in promoting sustainability of SMEs than more typical NGOs. As a result of this diversification of NGOs the struggle for leading positions in the network platforms and the competition for scarce funding and operate projects of international cooperation agencies have also intensified. This diversification of NGOs and, above all, the increasing of service-like NGOs aim to fulfill the business growth and market demands of SMEs in collaborating with market actors. Hence, new types of NGOs emerge to fulfill market demands.

The discourses that NGOs and SMEs endorse in the networks of sustainability of SMEs are: market adaptation, market access or market democratization in the first case study; business upgrading and corporate responsibility in the second case study; and cleaner production and appropriate technology in the third case study. NGOs and SMEs involved in the networks of organic production endorse one of the following three discourses: market adaptation, market access or market democratization. The main storyline of the first discourse is that NGOs and small scale producers are forced to get new capacities and to adapt to the free market. Small scale producers do not have the competences to adapt to the free market by themselves, and NGOs play a crucial role in assisting them. The main storyline of the second discourse is that small scale producers are eager to move to competitive markets. Support is needed from specialized agents in managerial and technological issues to organize supply to competitive local and international organic markets. The main storyline of the last discourse is the prioritization of making the organic market also interesting for low and medium income consumers. Rather than adapt or access to the free market, small scale producers intend to build up a fair relationship with the market by making organic products available to all income groups.

NGOs and SMEs in the business social responsibility networks endorse one of the following two discourses: business upgrading or corporate responsibility. In the first discourse, business social responsibility is seen as a strategy to match economic and social rights with sustainability of small scale enterprises. Connecting small scale enterprises with larger companies and influencing them to become sustainable is central in the discourse. In the second discourse, business social responsibility is seen as a business strategy that contributes to sustainability of larger companies and their supply value chains. Only small providers of large profitable value chains have the capacity to adopt social and environmental standards.

NGOs and SMEs involved in the networks of sustainable production endorse one of the following two discourses: cleaner production or appropriate technology. In the first discourse, cleaner production is seen as a business strategy to make SME production more efficient and sustainable. Allocating the most up-to-date modern technology is considered as the best way to reduce environmental impacts and increase competitiveness. The discourse focuses on SMEs that are well established in the local market and have the capacities to reach international markets. In the second discourse, appropriate technology is seen as tailor-made technology adjusted to the needs of SMEs, particularly of micro and small enterprises. Low capital, small scale and suitable technology for the local social, economic and cultural setting are central in the discourse. The discourse highlights the use of renewable energy, development of local markets and poverty fighting.

The seven discourses emphasize either market justice or sustainable market. This means that the discourses are different in their position towards social movement and the market. The discourses of sustainability of SMEs have evolved from long-standing antagonist discourses: the liberal market discourse on one hand and the social movement discourse on the other hand, which can be considered as the 'mother' discourses of the discourses of sustainability of SMEs. While the cleaner production discourse and the corporate responsibility discourse have their origins in the liberal market discourse, the market democratization discourse, the market adaptation discourse, the market access discourse, the business upgrading discourse and the appropriate technology discourse have their origins in the social movement discourse. Hence, the discourses of sustainability of SMEs share views with their mother discourses. Only, the market access discourse strongly diverges from its origins. The difference between market justice discourses and sustainable market discourses has to do with their interpretation of environmental reform and sustainability.

In sum, the identified changes are expressed in new roles for NGOs. Next to the usual 'watchdog' roles, NGOs are developing roles of 'helper' in order to answer to the market needs of SMEs. The new roles are performed not only by new types of NGOs but also by 'reoriented' conventional NGOs. Consequently, NGOs have become market agents as a result of their new roles. Finally, the findings contribute to the theoretical debates on network society theory and ecological modernization theory. The analysis of networks promoting sustainability of SMEs helps to understand more deeply the way non-state actors cooperate, and challenges Castells' scheme of space of flows versus space of place. Both spaces are connected and integrated in aiming for sustainability. Actors use rationalities, logics and power resources related to both spaces. Amending ecological modernization theory, the analysis suggests that it is needed to consider both ecological rationality and social rationality in order to advance environmental reform of SMEs in developing countries. The research also sheds light of issues of power. NGOs are becoming more collaborative and less confrontational, more conciliatory and less dogmatic towards market actors, but they remain rather conflictive and competitive towards fellow NGOs. Power of SMEs is not acknowledged in most discourses. However, SMEs show their power either by accepting or denying engage to the networks, either by collaborating or pressuring key actors and either by subscribing or being indifferent to the discourses. This power of SMEs pushes the networks to become more inclusive, participatory and valuable for SMEs. It rests on the capacity to be anchored within local social networks.

About the author

Walter Victor Castro Aponte was born on 6 June 1974 in the province of Huamalies, Huánuco, Peru. In 1999 he obtained the degree of biologist at the National University of San Marcos in Lima, Peru with a thesis on eco-toxicology applied to measuring water pollution, research conducted at the Lima Metropolitan Area Water Supply and Sewerage Service Company (SEDAPAL, in Spanish acronym). In 2002 he obtained the Ford Foundation Fellowship to start his MSc in Urban Environmental Management at Wageningen University and the Institute for Housing and Urban Studies (IHS), the Netherlands. He obtained his MSc degree in Urban Environmental Management with a major in Environmental Policy with the thesis on environmental management systems in Spanish local governments. In 2005 he started to develop an innovative PhD research proposal to be implemented in his home country Peru, involving Dutch and Peruvian NGOs and universities. In 2006 he obtained a Wageningen Sandwich PhD Fellowship to start his PhD study on Environmental Policy and to implement his PhD research proposal which ended up in this PhD thesis. Currently, he is working among others for the Instituto Latinoamericano de Ciencias (based in Peru) and the Instituto de Formación Ambiental (based in Spain) supervising and evaluating research theses, and tutoring post-degree students in urban environmental management, sustainable business and ecological agriculture topics at Latin American level using I&CT. His research interests focus on the linkages between environmental, social, cultural and political perspectives applied to environmental governance in Andean countries, particularly regarding the role of non-state actors.

Printed in the United States
by Baker & Taylor Publisher Services